Geological Society of America
Memoir 189

Phanerozoic Faunal and Floral Realms of the Earth: The Intercalary Relations of the Malvinokaffric and Gondwana Faunal Realms with the Tethyan Faunal Realm

Arthur A. Meyerhoff*

Arthur J. Boucot
Department of Zoology
Oregon State University
Corvallis, Oregon 97331

Donna Meyerhoff Hull
Programs in Geosciences
The University of Texas at Dallas
P.O. Box 830688
Richardson, Texas 75083-0688

and

J. M. Dickins
Australian Geological Survey Organization
P.O. Box 378
Canberra, A.C.T. 2601, Australia

1996

*Deceased.

Published by The Geological Society of America, Inc.
3300 Penrose Place, P.O. Box 9140, Boulder, Colorado 80301

Printed in U.S.A.

GSA Books Science Editor Abhijit Basu

Library of Congress Cataloging-in-Publication Data

Phanerozoic faunal and floral realms of the earth : the intercalary relations of the Malvinokaffric and Gondwana faunal realms with the Tethyan faunal realm / Arthur A. Meyerhoff . . . [et al.].
p. cm. -- (Memoir / Geological Society of America ; 189)
Includes bibliographical references and index.
ISBN 0-8137-1189-4
1. Paleobiogeography. 2. Plate tectonics. I. Meyerhoff, A. A. (Arthur Augustus), 1928- . II. Series: Memoir (Geological Society of America) ; 189.
QE721.2.P24P48 1996
560′.17--dc20 95-53781
CIP

10 9 8 7 6 5 4 3 2 1

Dedication

This work is dedicated to Professor Curt Teichert, a great scientist, paleontologist, and geologist, whose descriptions of thick marine sections in Western Australia and elsewhere revolutionized forever our perception and the concept of Gondwana. Photograph taken by John Talent in Buchan, Victoria, in 1976.

Contents

Foreword

This monumental study is the product of many years of investigations by scientists with worldwide experience, not only in field studies but also in assembling, integrating, and interpreting a voluminous suite of publications (more than 575 entries in English, French, German, Spanish, Russian, Japanese, and Chinese). These involve diverse disciplines such as paleontology, stratigraphy, structure, tectonics, geophysics, geochemistry, and others. Fully a third are not available in most scientific libraries; therefore, their existence has heretofore been unknown to many researchers. Curt Teichert, the internationally renowned scientist to whom this study is dedicated, was the inspiration for all the participants.

The prologue sets the tone for the present study, which is reminiscent of the classic essays of G. K. Gilbert and T. C. Chamberlin on scientific methods. This study might well be appropriately considered as an excellent modern day example of a "Study for Students" who are seriously interested in doing painstakingly accurate and thorough in-depth scientific research in these and allied disciplines. To test this idea, a group of outstanding young graduate students in the earth sciences was invited to study very carefully an advance copy of the manuscript. They became deeply disturbed, challenged, angry, and finally, excited. They commented: "If this global study of biodiversity through time is correct, and it is very convincingly presented, then a lot of what we are being taught about plate tectonics should more aptly be called 'Globaloney.' So—are we students deliberately taught only one part of a complex story? As a result, has our scientific objectivity become flawed? How will history judge the validity of our own future studies if we are not mindful of the message in the prologue of this paper?"

J. David Love
June 17, 1994

Geological Society of America
Memoir 189
1996

Phanerozoic Faunal and Floral Realms of the Earth: The Intercalary Relations of the Malvinokaffric and Gondwana Faunal Realms with the Tethyan Faunal Realm

ABSTRACT

Biogeographical data comprise a largely neglected but potentially powerful tool for deciphering the tectonic evolution of the Phanerozoic Earth. This is true because the borders of biogeographical realms, regions, provinces, and subprovinces are natural barriers, some of them tectonic in origin. Yet most major biogeographical realm boundaries, based on floral and faunal distributions, do not coincide with the partly tectonic, partly computer-generated boundaries of plate tectonics.

Instead, the paleontologic record shows that (1) a broad intercalary zone separates "northern" from "southern" biogeographical realms, and (2) this broad zone has existed during most, if not all, of Phanerozoic time. Within this intercalary zone, whose width ranges from several hundred to 5,000 km, strata bearing "northern" biotas are intercalated with strata bearing "southern" biotas and, in many areas, admixtures of "northern" and "southern" taxa are present within the same beds.

During the Middle and Late Cambrian, the southern realm is the Atlantic Realm. (A globally low Early Cambrian climatic gradient does not permit easy definition of a southern realm.) Following the Cambrian, the two principal southern realms are the Malvinokaffric Realm (Ordovician–early Middle Devonian) and its successor, the Gondwana Realm (Early Permian–Early Cretaceous). The two are separated in time by the later Devonian, an interval of rather cosmopolitan biotas.

In Asia and the southwestern Pacific, the boundary commonly used to separate the Malvinokaffric and Gondwana Realms from their northern equivalents (e.g., the Tethyan Realm in post-Paleozoic time) is the Taurus-Zagros-Indus-Yarlung suture zone and an implied eastward continuation to Papua New Guinea. For the Malvinokaffric Realm this boundary is not useful because the biota does not change across the suture. The boundary has been applied mainly to the younger Gondwana Realm for which it also is not useful. Gondwana elements (Lower Permian–Cretaceous) extend northward to the Tunguska basin of Siberia, to Mongolia, northeastern China, the Primor'ye region (north of Vladivostok), and the Kolyma River basin of northeastern Siberia. Conversely, northern—and especially Tethyan Realm—biotas extend southward to New Zealand, Western Australia, Northern Territory (Australia), southern India, and Saudi Arabia.

A suture zone is not present in Africa. Northern Africa, except for most of Ordovician and probably all of Silurian times, has been a tropical to subtropical region. During Early Permian through Tertiary times, Tethyan biotas are rather common in central Africa, less so but still present in central and southern Africa (e.g., Ghana, Niger, Republic of South Africa, Madagascar, Kenya, Tanzania, Ethiopia). A suture zone in Africa, if present, would have no real biogeographical significance.

Meyerhoff, A. A., Boucot, A. J., Meyerhoff Hull, D., and Dickins, J. M., 1996, Phanerozoic Faunal and Floral Realms of the Earth: The Intercalary Relations of the Malvinokaffric and Gondwana Faunal Realms with the Tethyan Faunal Realm: Boulder, Colorado, Geological Society of America Memoir 189.

A similar situation characterizes South America. In the Amazon basin the Silurian beds have Malvinokaffric Realm fauna. In the Amazon and Parnaiba basins during Early Devonian–Eifelian time, mixtures of Malvinokaffric and Eastern Americas Realms faunas are found, while from Mississippian-Pennsylvanian time onward, northern biotas dominate. Also, northern elements dominated the Pacific coastal zone at times as far south as southern Chile. Thus, as in Africa, a suture zone would, if present, have no meaning biogeographically.

Another problem also is evident from this study. As pointed out repeatedly by Teichert since the 1940s, marine conditions were widespread throughout the Gondwana Realm from Early Permian through Early Cretaceous times. Almost every marine embayment in rocks of these ages enters the *present* continent from the *present* shelf. This fact and the widespread presence of Tethyan taxa in significant parts of Gondwanaland suggest the presence there of a great deal of ocean water.

These are major problems and require successful integration into plate tectonics. Until the tectonics of the Earth and biogeographical data are integrated successfully, there will be no successful theory of tectonics.

PROLOGUE

As we understand science, its forward progress reflects the degree to which all relevant disciplines can be brought to bear successfully on a given problem. Thus biogeographers, paleontologists, stratigraphers, structural geologists, tectonophysicists, and geophysicists need to work together now to bring the powerful tool of biogeography into tectonic reconstructions and tectonic models. If plate tectonics is to have credibility, it must be consistent with biogeographical data.

STATEMENT OF THE PROBLEM

The problem that we address in this volume can be stated simply: the boundaries of the plates postulated by many geophysicists and geologists do not always match the boundaries of the biogeographical realms and lower rank units worked out by paleontologists and biostratigraphers. Nor do the proposed movements of continents correspond with the known, or necessary, migration routes and directions of biogeographical boundaries, either in time or in space. In most cases, the discrepancies are very large, and not even an approximate match can be claimed. We note that a majority of people working on global tectonic problems from the viewpoints of geophysics, tectonophysics, and structural geology disregard biogeographical data, or at best, treat them in an offhand manner. Those who mention such data make generalizations (e.g., "the biogeographical data support our model in a general way"), but rarely come to grips with significant details, many of which contradict their own conclusions. Their work, in fact, suggests no real familiarity with, or understanding of, either the facts or principles of biogeography.

Before we go any further we must ask; Should one expect biogeographical boundaries to correspond with plate boundaries? In biogeographical terms, the question is whether boundaries that prevent the reproductive communication responsible for biogeographical boundaries in the first place should correspond with plate boundaries? It is clear that in some cases the answer will be positive: i.e., the two types of boundaries should correspond. A positive example of this type is the separation of South American placental mammals, later Cretaceous to Pliocene, from those of North America. Thus, when plate boundaries correspond to boundaries preventing reproductive communication, the correlation is positive. But, as we show in this volume, many plate boundaries do not correspond with barriers to reproductive communication. Our extended treatment of India is a good case in point of a wholly negative correlation between plate boundaries and reproductive communication barriers. We need to ask why this concept of a 1:1 correlation in some areas between plate boundaries and biogeographical boundaries should have developed in the first place. It certainly was not due to the activities of most biogeographers. Rather, it appears to have been generated by physical scientists who assumed, probably owing to their lack of familiarity with biogeographical data, that separate plates, marine or nonmarine, some of which were situated in widely disparate parts of the globe, would perforce be inhabited by very different marine and nonmarine biotas.

In principle, scientists all agree that every piece of available evidence that could have any bearing on a question should be taken into consideration. In practice, however, one all too commonly finds that most scientists tend to restrict themselves to those classes of information with which they are reasonably familiar and comfortable. In terms of this treatment, there is a strong tendency for geophysicists to pay most attention to geophysical data, sedimentologists to lithological data, and paleontologists to paleontological data. For example, Jardine and McKenzie (1972, p. 20), writing near the beginning of the "plate tectonics epoch," stated: "It is no longer profitable for biologists to speculate about the past arrangement of land masses," a conclusion with which we strongly differ; we might almost phrase it the other way around for the pre-Carboniferous. In terms of paleogeographical maps on which biogeographical data has been plotted, most paleontologists have tended to use the most easily accessible or newest available map without worrying much whether their own information tends to support the reconstruction or is in any way consistent with the map. Since some of these maps rely entirely on geophysical data, it is not surprising that they are in apparent conflict with other classes of information.

Thus, contrary to widespread belief, hundreds, even thousands, of paleontological, stratigraphical, and related studies have proved that many taxa from Cambrian time to the present have paid little attention to postulated tectonic lines passing through the various continents. Noetling (1901), writing more than 90 years ago, observed that European faunas in Asia were underlain and overlain by Indian and other southern faunas and floras. This overlapping of realms is widespread from the western coast of South America to the eastern tip of Tethys, and is especially visible in strata of Carboniferous and Permian ages.

This situation raises important questions that earth scientists can no longer afford to ignore. For example, will geophysicists in general, and geologists who work in disciplines unrelated to biogeography, continue to disregard biogeographical data? Do those who briefly mention biogeography intend some day to come to grips with the serious problems and contradictions between their models and those of biogeography? As stated in the Prologue, all disciplines that relate to a problem must be brought to bear on that problem if the most reliable solution is to be obtained. In the case of tectonic models, we are convinced that to construct them without utilizing biogeography is rather like building automobiles for a peopleless planet. Of what use would they be with at least one essential component missing?

To illustrate the problem, we have chosen the broad geographical zone that separates the middle Paleozoic Malvinokaffric Realm—and subsequently the late Paleozoic and younger Gondwana Realm, the cool climate or southern realms—from warm realms of northern origin. This broad zone is transitional between northern and southern biogeographical realms; we call it the intercalary zone. Its width ranges from a few hundred to more than 5,000 km. Within it, some strata with northern taxa are intercalated with strata including southern taxa, and vice versa. In some places, northern and southern taxa appear within the same bed, or beds. With a single exception (which in fact may be no exception at all) the intercalary zone does not coincide with any of the plate boundaries postulated by geophysicists, structural geologists, and tectonophysicists. The single possible exception is the Indus-Yarlung suture zone north of India (Fig. 1), where many geoscientists claim that the suture zone does indeed coincide with a biogeographical boundary zone. What is puzzling is that such major inconsistencies between plate tectonic postulates and field data, involving as they do boundaries that extend for thousands of kilometers, are permitted to stand unnoticed, unacknowledged, and unstudied.

The fact that the inconsistencies have not been faced squarely implies to us that biogeographers, on the one side, and physicists and geologists, on the other, are not communicating. Yet the latter, by ignoring the biological disciplines, weaken their interpretations to the point of courting rejection (e.g., see Boucot and Gray, 1983), while ignoring a wealth of useful data that ultimately must be integrated with their own. In fact, the data from the fossil record comprise a powerful paleogeographical tool, one that must be reckoned with while resolving the problems of Earth dynamics. We believe, like Bucher (1964, p. 4), that "Ultimately the proof [of continental drift] must come from the geologic and palaeontologic record."

BIOGEOGRAPHICAL PRINCIPLES

The primary facts of biogeography are dictated by a few facts of astronomy. First, the collimated solar radiation reaching the Earth's surface decreases markedly per unit of area per unit of time away from the equatorial regions owing to a simple geometrical fact, namely, that the Earth is essentially spherical. Second, superimposed on this marked decrease in solar radiation per unit of area per unit of time from the equator to the poles is the obliquity of the Earth's axis of rotation and its orbit, which together produce the annual seasonality of the globe. The alternation of the seasons ensures that the equatorial regions will have a far more annually, unvarying environment (insofar as radiation is concerned) than the rest of the Earth. The biotas of the various latitudinal belts must possess adaptations permitting them to exist under conditions of either unvarying or regularly varying insolation, with all that both situations imply in terms of factors such as uniformity or regular variance in light, food supply, and temperature. We point out that the key characteristic of the tropics is not warmth or humidity, which can vary considerably in a tropical savanna or rain forest; rather, it is the relative constancy of conditions, that is, the comparatively low level of environmental fluctuations. In this sense, the frigid top of Mount Kilimanjaro is just as tropical as any tract of steaming rain forest in the Zaire (Congo) River drainage.

If one assumes that the Earth's obliquity has remained relatively constant during the last 600 m.y. or so for which we have an abundant fossil record, then it follows that a permanent, primary, tripartite, biogeographical subdivision (with gradation between adjacent parts) should be available wherever the record is available. The geological record provides a wealth of circumstantial data, some of which are summarized in this volume, in support of this permanent tripartite division. Biogeographical and lithofacies data of the Phanerozoic are consistent with this conclusion. We also point out that this primary tripartite biogeographical subdivision (high northern latitude, cool-cold; low latitude, mainly warm; and high southern latitude, cool-cold), with gradational units between them, is supported from the study of modern oceanic plankton (McGowan, 1971, Fig. 8A).

In addition to the latitudinal subdivisions dictated by distribution of solar radiation and the Earth's obliquity, it is necessary to consider the primary distribution of surface winds as well as their reactions with the surface waters of the oceans and epeiric seas. On land these factors may give rise to seasonally significant variations in rainfall, such as those that characterize the tropical savannas and the more seasonal rainforests.

Superimposed on these primary, latitudinal, biogeographi-

Figure 1. Index map, Eastern Hemisphere, to localities mentioned in the text.

cal divisions are longitudinally disposed barriers that break up the latitudinal divisions into additional biogeographical units. These longitudinal barriers are of many types. They include such things as landmasses, water masses with varied properties, submarine and other oceanic topographical features, and distance (which is a barrier for organisms unable to cross a certain environmental zone with ease). These longitudinally disposed barriers behave in a reciprocal manner for nonmarine and shallow marine organisms; that is, a break in a land barrier isolates nonmarine organisms present on either side of the break, but unites shallow water marine organisms formerly isolated from one another.

Involved also with these longitudinal barriers is the interaction among Coriolis force, surface wind–influenced oceanic currents, and differently disposed landmasses. It is safe to say that during no time interval of the past is there good evidence for a

EXPLANATION

EUROPE

1. BALTIC SEA
2. CAUCASUS RANGE
3. GERMANY
4. GREECE
5. HATTON-ROCKALL PLATEAU
6. MOSCOW BASIN
7. PARIS-LONDON BASIN
8. PAY KHOY
9. POLAR URALS
10. SPAIN
11. TAURUS MOUNTAINS
12. TURKEY
13. URALS

AFRICA

14. ACCRA EMBAYMENT
15. ALGERIA
16. BENIN
17. BOVE BASIN
18. CAPE PROVINCE
19. CHAD
20. CONGO RIVER
21. EGYPT
22. ETHIOPIA AND OGADEN DESERT
23. ETOSHA PAN
24. GABON
25. GHANA
26. GUINEA
27. GUINEA-BISSAU
28. KALAHARI BASIN
29. KAROO BASIN
30. KENYA
31. LESOTHO
32. LIBERIA
33. LIBYA
34. MAURITANIDES
35. MOROCCO
36. MOZAMBIQUE
37. MT. KILIMANJARO
38. NAMIBIA
39. NIGER
40. OKAWANGA BASIN
41. OMAN
42. OMAN MOUNTAINS
43. ROKELIDES
44. RUHUHU BASIN
45. SENEGAL
46. SIERRA LEONE
47. SOMALIA
48. SOUTH AFRICA
49. SUDAN
50. TANZANIA
51. TAOUDENI BASIN
52. TUNISIA
53. WARMBAD BASIN
54. WEST AFRICAN CRATON
55. ZAIRE
56. ZAMBIA
57. ZIMBABWE

ASIA

58. AFGHANISTAN
59. BORNEO
60. BURMA
61. HEILONGJIANG
62. INDUS-YARLUNG "SUTURE"
63. IRAN
64. IRAQ
65. ISRAEL
66. JUNGGAR BASIN
67. KAZAKHSTAN
68. KOLYMA RIVER BASIN
69. KUNLUN SHAN
70. KURUK TAGH
71. LHASA BLOCK
72. MALAY PENINSULA
73. LITIAN-JINSHA JIANG "SUTURE"
74. PAMIR RANGE
75. PERSIAN GULF
76. PRIMOR'YE
77. QINGHAI-XIZANG (TIBET) PLATEAU
78. SEA OF OKHOTSK
79. SHAANXI PROVINCE
80. SUMATRA
81. SYRIA
82. TARIM BASIN
83. THAILAND
84. TIMOR
85. TONKIN GULF
86. VERKHOYANSK RANGE
87. VIETNAM
88. WANDA SHAN
89. XINJIANG-UYGHUR AUTONOMOUS REGION
90. ZAGROS "SUTURE"

AUSTRALIA-NEW ZEALAND

91. BONAPARTE GULF
92. CARNARVON BASIN
93. MURIHIKU UNIT
94. NELSON AREA
95. NEW SOUTH WALES
96. NORTHERN TERRITORY
97. PAPUA-NEW GUINEA
98. PERTH BASIN
99. SOUTHLAND AREA
100. SYDNEY BASIN
101. TORLESSE ZONE
102. WESTERN AUSTRALIA
103. WESTRALIAN BASINS

perfectly latitudinal biotic distribution, owing to the complex interactions involved with the various and varied longitudinal barriers of the past. Currents, such as the warm Gulf Stream and cool Humboldt Current of the present, with their moderating influence both on land and sea, turn out to be normal features that perturb what might otherwise have been a far more latitudinally zoned biotic distribution pattern. However, the presence of such latitudinal departures in biogeographical boundaries is not meant to suggest that the overall, dominant latitudinal influence cannot be seen through the "noise" provided by the changing complexities of the longitudinal barriers.

In addition to these geographical features is the level of climatic differentiation present during the Phanerozoic. This level is ever changing, with some intervals characterized by a high degree of equitability and others, of which the present is an excellent example, characterized globally by a very highly differentiated climatic regime. The presence of Eocene crocodiles in northern Ellesmereland, earlier Cenozoic palms in west-central Greenland and in southern Alaska, and mangrove swamps in the early Eocene Paris basin makes this clear.

Therefore, Phanerozoic biogeographical history is essentially the record of organic evolution disposed against the fluctuating climatic picture and complicated by the geographical-topographical picture. Needless to say, no two intervals of Phanerozoic time have ever been characterized in detail by the same geographical and climatic history, although certain similarities may be drawn. In view of this fluid situation, it is not surprising to find that the superimposition of the Phanerozoic biogeographical units in detail does not show any consistent trend except the one imposed by the primary tripartite latitudinal seasonality caused by the Earth's obliquity and orbit.

REPRODUCTIVE COMMUNICATION

Biogeographical units are defined as areas, marine or nonmarine, that contain similar biotas. They may or may not be physically continuous; if they are not, then reproductive communication must be maintained. If reproductive communication is not maintained within a geologically short time interval, the reproductively isolated elements in the formerly uniform biota begin to evolve into separate taxa; i.e., the once biogeographically uniform biota becomes increasingly differentiated into unlike biotas.

Marine environment

Reproductive communication within the marine environment is maintained chiefly by means of larvae. Marine larvae consist of many types. Relatively sedentary forms, such as brooded larvae, or benthic larvae with limited powers of movement, tend to belong to highly provincial forms because they are unable to cross many types of marine barriers. A barrier for such a larval form would be any environment it was incapable of crossing for one reason or another. Conversely, planktonic larvae tend to have far greater powers of dispersal. Some planktonic larvae live for a time solely on the yolk supply, which may be large or small. The dispersal and reproductive communication possibilities of lecithotrophic larvae depend on the length of time granted by the yolk supply before metamorphosis to a benthic adult becomes mandatory.

Planktotrophic larvae are those that are able to feed on planktonic plants and/or animals, after exhausting their yolk supply, while in the plankton. They tend to have far greater powers of dispersal, and to spend longer time intervals as plankton before metamorphosis becomes necessary. A few planktotrophic larvae, the teleplanic types, are able to maintain themselves today long enough to permit the crossing of major oceanic water bodies, such as the tropical Atlantic. Today the percentage of planktotrophic and planktonic larvae is much greater in warm waters than in cool waters; in cool waters the percentage of planktonic larvae is very small. The implications of this information for the past are that extinct organisms dispersed by means of planktonic larvae would have had the capability of maintaining reproductive communication across significantly great water bodies from far-removed shallow water bottoms where the benthic adults flourished in reproductive isolation from each other, and that they were most likely associated with warm waters. Further, there is a complete spectrum in dispersability from the teleplanics to very localized forms that brood their own eggs.

Reproductive communication among planktonic protistans (e.g., foraminifera and radiolarians) is easily maintained within the oceanic water masses where they live. Endemism does exist, however, even among the most passive of planktonic organisms. The only likely barriers to reproductive communication among these organisms are physical barriers (e.g., continents, deep water sills) and the physical and chemical parameters of the oceanic water masses themselves (e.g., temperature, salinity, and nutrients).

Nonmarine environment

Reproductive communication in the nonmarine environment is more complex than in the marine. Higher land plants may be dispersed and maintain reproductive communication by means of spores (as in the bryophytes and lower tracheophytes) or pollen (as in the spermatophytes) that can be transported by winds. The higher tracheophytes that depend on seeds for dispersal tend to have more limited dispersal capabilities and be more endemic as a result. However, there are many exceptions to the above generalizations, and each group of land plants must be considered in terms of its own dispersal capabilities.

Nonmarine animals on land may disperse using a variety of devices. Tetrapods tend to disperse by walking, with varied barriers involving such things as large water bodies and unsuitable environments (e.g., deserts, montane regions, and harsh climates). The dispersal techniques employed by nonmarine invertebrates are legion, but they range from very efficient means of dispersal to ones that do not favor long-range dispersal and reproductive communication over great distances. Again, each group must be considered on its own merits.

A knowledge of the reproductive biology and dispersal capabilities of varied organismal groups should be considered before its reproductive communication capabilities can be evaluated reliably. *With wholly extinct organisms, however, a widespread dispersal pattern is prima facie evidence for the existence of an effective means of reproductive communication; conversely, a highly localized dispersal pattern shows that the means for reproductive communication are limited.*

Nonmarine vertebrate dispersal across marine barriers

It must be kept firmly in mind that the skins of modern amphibians are permeable to water, including sea water, which would upset their salt balance. Only a single, extinct Triassic amphibian group (Hammer, 1987) is associated closely enough with marine deposits to suggest that it had overcome this basic physiological barrier to dispersal. Mammals have such a high metabolic rate, and consequent water loss while breathing (John Ruben, oral communication, 1988), that they should be unable to cross lengthy salt-water barriers because they would become dehydrated in the process. Only terrestrial reptiles have the capability to disperse more regularly across lengthy salt-water barriers, as G. G. Simpson noted (*in* Meyerhoff and Meyerhoff, 1974) . Again, an understanding of the biology of each group of organisms must be considered before their dispersal and reproductive communication capabilities can be understood. Thus it is logical for varied Mediterranean region mammals to have dispersed to relatively nearby islands during the Pleistocene, even in the absence of land connections, whereas the absence of mammals on oceanic islands like the Galapagos (Fig. 2) makes good sense in terms of the immense distances involved, as does the virtual isolation of Madagascar from later Cenozoic mammalian invasions from the African mainland (Fig. 1).

EXAMPLES OF THE PROBLEM

The literature on faunal and floral realms is vast and somewhat bewildering to anyone unfamiliar with the biological and paleontological records. Literature on Malvinokaffric and Gondwana faunas and floras, on the one hand, and Eastern Americas, Old World, Tethyan, Angara, Cathaysian, and Boreal (to name a few) faunas and floras, on the other hand, would lead the casual reader to conclude—incorrectly—that there is no conflict between biogeographical and plate boundaries, except in southern and southwestern Asia, where it is only a matter of determining which of several interpreted sutures was the boundary between southern and northern realms at different times in the past, and/or labeling anomalous areas as displaced terranes. No conclusion could be more wrong. Some prominent examples are mentioned below.

Americas

Plate models require that a major suture separate North from South America both north and south of the intervening Caribbean plate. Later Paleozoic biogeographical data, however, show that the break between northern and the southern marine Malvinokaffric and Gondwana realms parallels the Amazon Valley, lying at times north, at other times south, and sometimes

Figure 2. Index map, Western Hemisphere, to localities mentioned in the text.

within the valley (Fig. 2). Tectonically, this valley corresponds to the transcontinental Huancabamba fracture zone which has not been active since late Proterozoic time (Fig. 2) (Ham and Herrera, 1963; de Loczy, 1970). Thus, in plate tectonic terms, most of South America north of the Amazon River, an area of 3,300,000 km^2, has belonged biogeographically to the North American plate since Ordovician time.

Boucot and Gray (1979) noted that an additional boundary in South America strikes north-south ("Andean" boundary), separating a warm-water area on the west from cooler continental areas on the east. The north-south boundary—except for Cambrian and Ordovician times—bends eastward through the Amazon Valley. During the Cambrian and Ordovician, the cool-warm boundary passed north of South America. To be consistent, plate tectonic models should provide for a major north-south 7,500-km-long (4,650 mi) suture within or close to the modern Andes.

Africa-Arabia

Plate models require a suture zone running the length of the Mediterranean Sea, despite stratigraphic continuity between Europe and Africa. Biogeographical data, however, beginning in Devonian time, show a break only *within* the African continent, extending from near Dakar to Arabia (Fig. 1). (In pre-Devonian time, the warm-cool boundary was in Europe and/or the Mediterranean.) Here, too, the area of faunal and floral overlap is vast—8,200,000 km^2 (3,150,000 mi^2) north of a line between Dakar and Saudi Arabia.

Australia

Although Australia (Fig. 1) is proposed to be a part of a plate well south of any relevant suture zone (e.g., Drewry et al., 1974), the fact is that much of northern and western Australia is in a northern or warm Tethyan region. As Teichert (1958) has pointed out, marked differences separated eastern and western Australia during several intervals of time so that, if plate tectonic precepts are applied here, a suture should separate eastern from western Australia. Again, a huge area is involved, more than 3,000,000 km^2 (1,860,000 mi).

Indus-Yarlung suture zone

This proposed suture, extending more than 5,000 km (3,100 mi) across south-central and southeastern Asia alone, is projected westward to the Taurus Mountains of Turkey and the Troodos massif of Cyprus; it is projected southeastward through Papua New Guinea (Fig. 1). In many plate tectonic models, it commonly is the boundary between northern and southern plates. Projection of this logic to biogeography leads to the conclusion that the suture zone also separates northern from southern biogeographical realms. Because of the logic of this plate tectonic reasoning, is it not also appropriate to apply this same reasoning to the Americas, Africa, and Australia, as discussed above? Through an examination of the Indus-Yarlung suture zone, let us see whether the logic holds.

As long as Gondwanan taxa were known only from south of the suture zone, the argument of plate tectonics was consistent. However, when Gondwanan taxa were found in northwestern China (Xinjiang), Manchuria (Heilongjiang Province), the Primor'ye region north of Vladivostok, the Tunguska basin northwest of Lake Baykal, and the Kolyma River basin along the Arctic Siberian coast (Fig. 1), their identifications were ignored; many were said to be misidentifications, even when careful documentation was presented (e.g., Zimina, 1967; Samylina and Yefimova, 1968; Sun Ailin, 1973a; Fang et al., 1979; Kalandadze and Rautian, 1983; Zhang Lujin, 1983a,b; Gu Zhiwei et al., 1984). Yet the reality of most of the identifications, such as the bones of the Triassic reptile *Lystrosaurus* from Xinjiang, cannot be denied, because most of the fossils in question have been examined by experts (Vozenin-Serra, 1984).

Once some of the Gondwana taxa from north of the Indus-Yarlung suture zone were accepted, the collision zone was shifted northward to embrace larger and larger segments of Asia (e.g., Crawford, 1974; Stauffer, 1983). Gondwana's preeminence over Tethys is curious. Tethyan taxa are known from many places on the Indian craton and other places south of the suture, yet the suture's position is always shifted to accommodate Gondwanan taxa, never the reverse! When Tethyan forms are found south of the suture, they are said to mark the northern shore of Gondwanaland. This leads to speculation on the fate of the southern shore of Angaraland, yet this point is rarely discussed in the literature.

Anyone who studies the pre-Gondwana Realm strata (fauna and lithology) of the Himalaya-Xizang (Tibet) region is struck at once by the fact that warm water taxa occur on both sides of the suture (Mu Enzhi et al., 1986). This raises the question of why a suture should have formed in southern Xizang (Tibet) in the first place. The region has almost always been a stable craton (e.g., Huang and Chen, 1987; Taner and Meyerhoff, 1990) during the Phanerozoic and no a priori, obvious reason exists to explain why the Indus-Yarlung suture (or, for that matter, other postulated sutures: e.g., Sengör, 1984) should ever have developed here. This problem has never been answered satisfactorily.

From the foregoing, it is evident that many questions remain unanswered. However, to enable the reader to ask the appropriate questions, we present some of the problems in greater detail.

BIOGEOGRAPHICAL TERMINOLOGY

Biogeographers employ a hierarchy (from largest to smallest): realm, region, province, and subprovince. The largest units—of which there are rarely more than three or four for each interval of time—are more mutually exclusive in terms of the taxa they contain than are the smaller units. For example, modern Arctic and Antarctic Realm birds are largely exclusive: there are no penguins in the north, and there are no alcids (auks, guillemots, or puffins) in the south; hence, the birds characterize two distinct realms. Similarly, the Old World and New World Realms of tropical plants and animals are distinctive and are very different at many taxonomic levels; both represent distinct realms. In the first case (that of the polar birds), reproductive isolation is maintained by the low-latitude tropical-subtropical belt separating them. In the second case (that of the two tropical realms), reproductive isolation is maintained by the sheer width of the low-latitude, isolating oceans.

Attempts have been made to "quantify" the definitions of realm, region, province, and subprovince, but no totally satisfactory system has yet been devised. For example, the Arctic and Antarctic share very few families, genera, or species. Their really unique character involves the very large number of higher taxa that are absent from both (i.e., major groups that have never adapted themselves to the rigorous polar conditions of the Neogene-Quaternary).

Because of carelessness or misunderstanding, several items have been misused over time. Therefore, we define these terms as used in our text.

ATLANTIC REALM

Descriptions of the Atlantic, Malvinokaffric, and Gondwana Realms are summarized here, and followed by a more detailed treatment of the Gondwana Realm, especially as the latter relates to northern realms, mainly the Tethyan.

The term "Atlantic Province" was used by Walcott (1889) to characterize the Middle and Upper Cambrian trilobites of much of northern Europe and the eastern fringe of North America (eastern Newfoundland, coastal New Brunswick, Nova Scotia, eastern Massachusetts, Rhode Island, the Carolina Slate Belt of eastern Virginia, the Carolinas, and Georgia; Fig. 2). Coeval but distinctly different Cambrian trilobites west of eastern, coastal North America were placed by Walcott (1889) in a Pacific Province (in 1889, the biogeographical heirarchy extending from realm through subprovince had not been formalized). We now recognize that the Atlantic Realm represents cool climatic conditions as deduced from lithofacies and biofacies evidence of the types discussed below. The Pacific Realm represents warmer climatic conditions.

The Middle and Upper Cambrian Atlantic Realm is succeeded by a similar Ordovician litho- and biofacies that ultimately gives way to the environmentally similar Malvinokaffric Realm. Possibly the best place to put the break between the younger Malvinokaffric Realm and the older Atlantic Realm is at the Ibexian-Whiterockian (=Canadian-Mohawkian of older North American nomenclature; Fig. 3) boundary where there is a distinctive extinction event followed by a major global adaptive radiation.

MALVINOKAFFRIC REALM

The term "Malvinokaffric Realm" was introduced by Richter (1941) to replace the terms "*Flabellites* Land" and "Austral province" introduced by J. M. Clarke (1913). The term was used originally for the highly endemic, Devonian, marine, benthic, invertebrate faunas occurring in sedimentary sequences showing evidence of deposition in a cool to cold environment in South America, the Malvinas (Falklands) Islands, and South Africa. We understand now that the Devonian age range of the faunas showing these characteristics—and despite correlation difficulties (no goniatites, no conodonts)—is approximately Emsian-Eifelian (Fig. 3) (Boucot, 1985, 1988; Melo, 1988). Boucot (1975, 1988) has pointed out the largely Eastern Americas Realm origins (that biogeographical unit occupying much of eastern North America and northern South America) of many of the taxa. The term has been extended to include the highly endemic Silurian faunas characteristic of the same general region, and even some of the similar later Ordovician faunas. It was this same region, plus some adjacent areas, that were subsequently occupied by the Gondwana Realm.

There is no evolutionary continuity between the Silurian Malvinokaffric taxa and those of the Devonian, owing to an intervening extinction event. The younger extinction event that affected the Devonian Malvinokaffric taxa is now attributed to a global warming trend (Boucot, 1988), but the cause(s) of the terminal Silurian event is (are) unknown. A minor extinction event is recognized elsewhere in this volume at the end of the Silurian (Boucot, 1985), or possibly within the Gedinnian (Fig. 3), but its cause also is uncertain. The origins of the Silurian Malvinokaffric fauna are uncertain, as are the origins of the Ordovician Malvinokaffric forms (which became extinct near the end of the Ordovician).

Within the Malvinokaffric Realm (spelled "Malvinocaffrisch" by Richter, 1941), the lithofacies consist of dark-colored terrigenous clastics with obvious unweathered, detrital mica flakes. Conspicuous by their absence are warm-water indicators such as carbonates, reefs, evaporites, redbeds, laterites, and bauxites.

Overall, the Malvinokaffric Realm faunas, like those of the Neogene-Quaternary polar marine benthos (Fischer, 1960; Arctic and Antarctic) and plankton, are of low diversity, from the class through the species levels (i.e., total numbers of known taxa within the realm). The Malvinokaffric taxa, like those of the polar Neogene-Quaternary, occur in low-diversity communities that contrast markedly with coeval communities that lived in warmer climates. The absence of warm-water indicators—the reef complex, bryozoan thickets, sponge forests, pelmatozoan thickets, and other communities—is notable. In fact, in many ways, the trademark of the Malvinokaffric fauna is the *absence* of certain major groups (most corals, conodonts, stromatoporoids, nautiloids, calcareous algae, most bryozoans) and the *abundance* of conularids and hyolithids. Unlike the warmer climate communities, a Malvinokaffric Realm fauna is not described in terms of the percentages of this group, that group, and another group, all of which are likely to be present. The total number of classes, orders, families, genera, and species is small. The fauna and lithofacies together suggest cool to cold climate, and a fair degree of geographical isolation. Malvinokaffric faunas, although concentrated in southern South America, southern Africa, and Antarctica, did extend at times across broader areas, even reaching into North Africa, Arabia, Western Europe, and a limited portion of easternmost North America.

Figure 3. Time scale illustrating the terminology used in the text. Radiometric dates are from Harland et al. (1990).

The origins of the Malvinokaffric Realm biotas of Ordovician age are poorly known. This lack of knowledge may reflect largely the limited state of our understanding of overall Ordovician biogeography. Biotas of Silurian age similarly are of uncertain origins, although they are very distinct from extra-Malvinokaffric Realm taxa. The Devonian Malvinokaffric Realm faunas, after total extinction of the Silurian Malvinokaffric fauna by at least Gedinnian time (Fig. 3), were derived in largest part from Eastern Americas Realm faunas (Boucot, 1975, 1988). The Devonian fauna became extinct near the end of the Eifelian (Boucot, 1988; Melo, 1988).

GONDWANA REALM

Introduction

The concept of Gondwana-Land was introduced by Suess (1885, 1888). In its type region, India, the Gondwana biogeographical realm usually encompasses the time interval from Early Permian through Early Cretaceous. A history of the Gondwanaland concept was published by Teichert (1952). The concept rests historically on two perceptions (Fairbridge, 1965; Teichert, 1974): the widespread distribution of continental sediments containing the *Glossopteris* flora, and the absence of marine strata of late Paleozoic–early Mesozoic age from the Southern Hemisphere and Peninsular India; and a uniform distribution of glacigene rocks of late Paleozoic age in the same area. As originally conceived, the lands included in Gondwanaland proper were characterized by a near-total dominance of nonmarine continental sedimentary deposits, and by a characteristic sequence of terrestrial floras and vertebrate faunas. Because marine beds were believed to be so scarce, Gondwana sequences were not easily correlated with coeval European, North American, and Asian sections. But as more marine tongues were discovered, correlations around the world steadily improved. Even so, substantial disagreement exists in some regions, particularly where intra-Gondwana correlations are involved.

The discovery of thick marine sequences (as in Western Australia and South Africa) has discredited the original Gondwanaland concept to a large degree (see Teichert, 1939, 1940, 1941, 1942, 1943a,b, 1949, 1950, 1951, 1958, 1972, 1974). Consequently, we use Teichert's (1974, p. 361) definition of the term *Gondwana region*, equating it generally with the biogeographic term *Gondwana Realm*. The Gondwana region thus comprises ". . . the greater part of Argentina and Brazil, southern and central Africa, Peninsular India, Australia, the intervening island complexes of the Falkland Islands and Madagascar, and Antarctica. Marginally included are such areas as the Precordillera of western Argentina, the Amazon trough, the Salt Range of Pakistan, New Guinea, and New Zealand—because some of these have been sources of marine faunas that invaded the Gondwana region." These lands include most areas of the older Malvinokaffric Realm plus outlying regions (e.g., Australia, Madagascar, India), where older Malvinokaffric Realm faunas are unknown. To the list of marginal areas can now be added the Qinghai-Xizang (Tibet) Plateau area of China, together with regions west and southeast of the plateau. In fact, as geological knowledge is accumulated on the faunas and floras of Central Asia, China, and the Russian Far East, surprising outliers of Gondwana are being discovered, far removed from the original Gondwanaland area.

Our primary interest here, however, is to examine the intercalary relations that exist between the northern realms and the Gondwana Realm, as well as between the northern realms and the Gondwana-allied, older Malvinokaffric Realm. Consideration of the still older Cambrian Atlantic Realm relations is also useful.

Most of the marine faunas now known from the Gondwana Realm, like their Malvinokaffric Realm predecessors, are believed to be cold-water faunas. The age of these cold-water faunas (called *inter alia* Gondwana fauna and *Eurydesma* fauna), of the associated *Glossopteris-Gangamopteris* flora and microflora, and of the intimately related glacial deposits regarded as characteristic for the Gondwana region has been the cause of much contention and confusion. This partly reflects the difficulties of correlation because of the distinctive provincialism produced in part by strong climatic differentiation. The contention and confusion also partly reflect an inadequate appreciation, or a misunderstanding, of the paleontological information that is available.

The age of the cold-water Gondwana biota and deposits has been discussed by authors too numerous to mention in this volume. The best review of the age is available in Archbold (1982), who concluded, on very substantial grounds, that the Gondwana Realm faunas and the associated floras and microfloras are (p. 267) ". . . Latest Asselian or younger in age, and that most of the underlying glacial beds are probably Early Permian (Asselian) in age." This conclusion was further strengthened by Dickins (1985a). Hence, in the present work we regard the Asselian as the earliest stage of the Permian (Fig. 3) (see also Archbold, 1991a) and the beginning of the most extensive Gondwana glaciation.

Palynological data (Kemp et al., 1977; Truswell, 1980) indicate that the most extensive glaciation (the main glaciation of Dickins, 1985a) is coeval in the various parts of the Gondwana region, although Kemp et al. (1977) regarded the time of the main glaciation as Late Carboniferous rather than Early Permian. The grounds of Kemp et al. (1977) for the Late Carboniferous age assignment are very weak, as Archbold (1982) has shown. Strong support for the Asselian age of the oldest glacigene strata has come from a palynological study by C. B. Foster (*in* Wopfner and Kreuser, 1986; see also Wopfner, 1991). Foster, studying the Karoo sequence of the Ruhuhu basin along the shore of Lake Nyasa (Lake Malawi), southwestern Tanzania (Fig. 1), found a typical Asselian microflora in the glacigene beds. Marqués Toigo (1991) came to a similar conclusion for the age of a microflora in the basal tillite sequence of the Brazilian part of the Paraná basin.

For a relatively recent review of the glaciation, readers are referred to Dickins (1985a). Regarding the ages, we follow Archbold (1982) and Dickins (1985a). Accordingly, the earliest Gondwana marine fauna in Peninsular India is latest Asselian or even early Sakmarian (Tastubian; Fig. 3). Many authors refer to this time interval as latest Carboniferous–Early Permian, an age assignment that recognizes the broadest time interval possible for cases where definite information is unavailable (e.g., Clarke, 1990). However, we emphasize that where such data are available, they show only an Early Permian age.

The identification of glacial and glacial marine beds, and the determination of their ages, are only two of a host of problems that beset the Gondwana-Tethys concept. Together with many other problems, they have been reviewed by Meyerhoff and Teichert (1971), Teichert and Meyerhoff (1972), Meyerhoff and Meyerhoff (1978), Boucot and Gray (1979), Haller (1979), Auden (1981), Dickins and Shah (1981), Gansser (1981, 1991), Waterhouse (1983), Stöcklin (1984), Dickins (1985a,b), Chatterjee and Hotton (1986), Saxena et al. (1986), Huang Jiqing and Chen Bingwei (1987), and Zheng Haixiang and Zhang Xuanyang (1990). The problems raised by these authors must be resolved if a satisfactory solution is to be found. The recent book by Huang Jiqing and Chen Bingwei (1987) is of special interest because of the light it sheds on the many problems that Tethys and Gondwana face within China, especially as new data are gleaned from the more remote areas of that country. These many problems are unlikely to be solved by the solutions of some authors (e.g., Dewey, 1988), because their models commonly break up contiguous parts of a single faunal realm, region, or province and scatter these parts onto separate landmasses thousands or hundreds of kilometers apart in widely different climatic zones, evidence for which we document subsequently. To illustrate this point, we have presented the data for the Devonian and Jurassic on two separate maps—one showing the present-day distribution of oceans and continents, the second using the paleogeographical reconstructions of Drewry et al. (1974; cf. Fig. 7a with 7b and Fig. 12a with 12b). Although the paleogeography of the former maps may not be perfectly consistent with the time periods that they represent, overall distributions of floral and faunal realms, as well as climatically influenced deposits, are far more compatible with these maps than with the reconstructions of Drewry et al. (1974).

What kind of a world was Gondwanaland? The published statements relating to this question reveal at once a lack of agreement on this topic. Of some importance is the fact that, as more data are collected from the faunal and stratigraphical records, the diversity of opinions increases. The full spectrum of Gondwana concepts ranges from that of a single large supercontinent to one of dispersed continents arranged much as they are today.

Marine invertebrates of the late Paleozoic

Until World War II, many areas of Gondwana were almost unexplored. One of the principal explorers in the region was Curt Teichert, whose descriptions of thick marine sections in Western Australia and elsewhere revolutionized forever our perception, and the concept, of Gondwana (Teichert 1939, 1940, 1941, 1942, 1943a,b, 1949, 1950, 1951, 1952, 1958, 1967, 1972, 1974, 1991; Teichert and Rilett, 1974).

During Mississippian time, the Tethys Sea extended from the western Mediterranean region to Papua New Guinea and northwestern Australia (Mamet and Belford, 1968; Mamet and Playford, 1968; Mamet, 1972, 1973). Elsewhere, marine sequences are moderately abundant along the Atlantic coasts of South America and Africa. They are known from offshore West Africa (Kamen-Kaye and Meyerhoff, 1979; Martin, 1982; Meyerhoff, 1984), northeastern Brazil (Petri and Fulfaro, 1983), and southern Argentina (Harrington, 1962). Probable Mississippian is present in South Africa (Theron, 1962; Gardiner, 1969; Teichert, 1974; Loock and Visser, 1985). Mississippian faunas of North American aspect are found in the Precordillera of western Argentina (Mesigos, 1953). Australian faunas have affinities with both Western Europe and Midcontinent America. Some forms are conspecific with those found in Europe (Roberts, 1985).

Overall, the Mississippian faunas are more cosmopolitan, less provincial than the older Devonian (pre-Frasnian) and younger Pennsylvanian-Permian faunas. Hill (1973, p. 14), viewing the Mississippian coral fauna, stated that, "In my view the presently known arrangement of the coral zoogeographic regions, provinces and sub-provinces of the world in the Lower Carboniferous epoch fits well with a distribution of the continents about the Polar Sea very similar to that of today. . . ." Australia's near-isolation, as indicated from coral studies, is striking. Mamet and Belford (1968) reached a similar conclusion from foraminiferal data, but commented on northern Australia's close relations with Tethys. Young (1987), working with Devonian vertebrates, also commented that present geography, not Pangaeic geography, explains Devonian distributions best, although the vertebrate sample is far more limited during this interval than is true for the invertebrates.

Gondwana faunas and floras made their appearances during Pennsylvanian and Permian times. C. A. Ross (1973, 1979), on the basis of fusulinid foraminifers, recognized two provinces: a Eurasian-Arctic Province (North Africa, Europe, Asia, northern Australia, northwestern North America), and a Midcontinent-Andean Province (central and southwestern North America, western South America, Amazon basin), with a broad transition zone between them. By late Early Permian time, these had differentiated further into three regions: Tethyan-Western Cordilleran, Uralian-Franklinian, and Midcontinent-Southwestern North American (including northern South America). The Gondwana area forms a fourth distinct unit, because it is devoid of the groups used by Ross to delineate his "provinces" and "regions."

Gobbett (1973a) arrived at a similar but simpler paleozoogeographical scheme on the basis of fusulinids. He recognized distinct Boreal, Tethys (including northern South America, northwestern Australia), and Gondwana Realms.

Stehli (1973, and many other publications) published several studies showing taxic-diversity gradients for Pennsylvanian-Permian time. For this purpose, he used verbeekinid fusulinids; waagenophyllid, durhaminid, and lophophyllid corals; and dasycladacean algae. He found that species diversity increases markedly toward the present geographical equator, and concluded that the geographical North Pole for Pennsylvanian-Permian time was about the same as that today.

Ustritskiy's (1967, 1973) and Ustritskiy and Stepanov's (1976) studies of Russian Permian invertebrates strongly support Stehli's work. They found similar taxic diversity gradients across Eurasia, and placed the Permian North Pole close to the present Lena Delta (Fig. 1). An impoverished Permian fauna, attributed to a cold climate, is present around Ustritskiy's (1973) pole, forming a Boreal Realm that coincides rather well with the modern faunally impoverished Arctic region. A somewhat similar conclusion was reached by Dutro and Saldukas (1973). Dagis and Ustritskiy (1973) noted that the Permian Boreal Realm is complemented by an antipodal Permian Austral Realm. Diamictites, common in the Permian Austral (Gondwana) Realm, also seem to be widespread in the northern part of modern-day Russia. We review briefly the evidence for glaciation-related diamictites in the Commonwealth of Independent States (CIS), particularly in European and Asiatic Russia.

Comments on late Paleozoic glaciomarine deposits of northern Russia

The precise nature and ages of the late Paleozoic diamictites of northern Russia have not yet been resolved. Most workers in the region believe that tillite, if present, is restricted in areal extent, and that most of the pebbly mudstones described from the field are ice-rafted deposits. Boulders as large as 2 m (6.6 ft) have been observed being carried out to sea from the shoreline region of Labrador (Rosen, 1979), which is consistent in the following Russian examples with transport from a rocky shoreline, as well as with the incorporation of some blocks of soft sediment. At least two glaciomarine zones have been recognized in northern Russia and one more is thought to be present by some.

The oldest pebbly mudstone unit is Pennsylvanian, lying at or near the Bashkirian-Moscovian boundary (Ustritskiy and Yavshits, 1971; Epshteyn, 1981a) (Fig. 3). Ustritskiy and Yavshits (1971) cited a gray to dark gray mudstone-siltstone matrix with 15 to 30% clastic material (sand size and larger) that includes some 30-cm (12-in) boulders; average pebble size is 1 to 3 cm (0.3 to 1 in). The clasts include rhyolite, rhyolite porphyry, quartz keratophyre, vein quartz, and Bashkirian limestone (the largest limestone boulder found was 30 × 25 × 20 cm [12 × 10 × 8 in]) from the Kolyma River area. This zone occurs in a 500-km-long [310 mi] band that extends from near the Popovka River (64°30'N, 151°30'E) in the southwest to near the Berezovka River (66°15'N, 158°30'E) in the northeast. The associated fauna includes the brachiopods *Krotovia mirabilis*, *Balakhonia* aff. *settedabanica*, *Fluctuaria* ex. gr. *cancriniformis*, *Orulgania* cf. *gunbiniana*, and *Taimyrella pseudodarwini*, together with the ammonoids *Parajakutoceras secretum* and *Bisatoceras* (Ustritskiy and Yavshits, 1971). This pebbly mudstone unit, therefore, appears to be younger than the Carboniferous glaciation in western Argentina and southern Bolivia (Harrington, 1962). Gonzalez (1990) dated the Mississippian glaciation in South America as late Viséan–early Namurian (early Serpukhovian: Fig. 3). However, the Russian and South American glaciations may have occurred during the same cool interval.

The youngest pebbly mudstones interpreted to be glaciomarine are Kazanian (Late Permian; Ustritskiy and Stepanov, 1976; Epshteyn, 1981b) (Fig. 3). This zone is geographically widespread, occupying an area of at least 550,000 km^2 (212,000 mi^2) that includes the Verkhoyansk Range from the Arctic Ocean to the Sea of Okhotsk, as well as an extensive belt paralleling the entire northern shore of the Sea of Okhotsk (Fig. 1). The fauna includes the brachiopods *Cancrinelloides obrutshewi*, *Strophalosia* ex gr. *sibirica*, *Neospirifer invisus*, and others that confirm the Kazanian age (Mikhaylov et al., 1970). Ustritskiy (1973) mentioned boulders as large as 1.5 m (5 ft).

A possible third pebbly mudstone zone, also of Permian age, was described by Andrianov (1966). According to him, and to Ustritskiy (1973), this pebbly mudstone sequence is of Sakmarian age (Early Permian). It occupies a large area of the Verkhoyansk Range and parts of Novaya Zemlya, Pay Khoy, and the Polar Urals (Ustritskiy, 1971, 1973). Among the marine taxa is the brachiopod species *Jakutoproductus cheraskovi* of the *J. verkhoyanicus* group. According to Betekhtina and Bogush (1984), the *J. verkhoyanicus* group, including *J. cheraskovi*, does not range below the Asselian (Fig. 3), and its lowest occurrence in the section is taken as the base of the Permian in the region. It is a characteristic genus of the Permian Boreal Realm as defined by Ustritskiy (1971, 1973). However, in a later publication, Ustritskiy and Stepanov (1976), wrote that, despite the presence of *Jakutoproductus cheraskovi*, this third pebbly mudstone zone was really equivalent to the second, or Kazanian, zone. They stated that the lithological details of both are identical and therefore that the two must be equivalent.

Regardless of the final outcome, Ustritskiy (1971, 1973), Ustritskiy and Yavshits (1971), Dagis and Ustritskiy (1973), Ustritskiy and Stepanov (1976), Betekhtina and Bogush (1984), and many others have demonstrated that, beginning in Bashkirian-Moscovian time, a cooling trend affected the present Arctic region and is clearly reflected in the fauna and flora. According to Dagis and Ustritskiy (1973) and Ustritskiy (1973), the faunas of the Permian Boreal Realm are, like the faunas of today from the same region, impoverished, with very few taxa. In fact, today's Boreal Realm and that of the Permian coincide very closely according to the following reasoning: (1) the number of taxa is greatly reduced; (2) there are no fusulinids or colonial corals; (3) moreover, the woody plants of the Permian Boreal Realm have well-developed tree rings (coeval woody plants of the Tethyan Realm farther to the south have no rings); and (4) therefore the conclusion that the Earth, during Permian time, had a bipolar biotal distribution similar to that of today seems to be inescapable. We return to the topic of bipolarity in a subsequent section.

Triassic-Jurassic invertebrates of Gondwana

Kummel (1969) and Khudoley (1974), studied, respectively, the Scythian Stage and all Jurassic stages (Fig. 3). Their work, based on ammonoid distributions, led them to identical conclusions: their data encounter serious problems with Pangaeic reconstructions. Khudoley and Prosorovskaya (1985) came to the same conclusion. Kummel (1969) and Khudoley and Prosorovskaya (1985) also constructed taxic diversity gradients. Like Stehli (1973) and Ustritskiy (1973), they found that the isodiversity lines are almost symmetrical with respect to the present geographical pole.

Biogeographical studies, such as those conducted by Ustritskiy (1967, 1973), Kummel (1969), Stehli (1973), Khudoley (1974), and Khudoley and Prosorovskaya (1985), support Waterhouse's (1983) and Stöcklin's (1984) contention (the latter on stratigraphic and structural grounds) that the wide Tethys postulated by Dietz and Holden (1970), Drewry et al. (1974), and Johnson et al. (1976) never existed. Many other stratigraphical and structural studies conducted in recent years in Asia has come to the same conclusion (e.g., Gansser, 1981; Foster et al., 1994), or at the very least indicate that great caution is necessary (Kamen-Kaye, 1972). We return to this problem in succeeding sections.

Tetrapods

Charig (1971, p. 126) wrote, after a thorough review of Permian and Mesozoic tetrapod faunas, that "the Tethys [Sea] itself could have been no more than a minor obstacle to the north-south migration of land animals." He stated further that "the evidence is sufficient to show that there was a cosmopolitan distribution of families—in some cases of genera—through the whole of the [Mesozoic] era." In varying degrees, nearly all vertebrate paleontologists whose specialties are late Paleozoic–Mesozoic tetrapods are in agreement with Charig (e.g., Cox, 1973; Kalandadze, 1974; Romer, 1975; Colbert, 1979, 1981; Cosgriff and Hammer, 1983; Kalandadze and Rautian, 1983; Gao Keqin, 1989). Implicit in Charig's (1971) interpretation is the absence of a deep, broad ocean basin on the site of the postulated Tethys. The likelihood that Charig (1971) is correct is illustrated by the computer-derived map of Cosgriff et al. (1982) on which those authors show the distribution of the Early Triassic reptile *Lystrosaurus*. That distribution (Antarctica, Africa, Russia, India, China, Vietnam) is much easier to explain in terms of modern geography than of the Pangaeic map because of the shorter distances that *Lystrosaurus* would have had to travel had the continents not shifted. Although future collecting may find *Lystrosaurus* on the continents on which it is now unknown (the Americas, Australia), its absence from Australia, at least, is easier to understand on the basis of today's geography.

Australia's relative isolation in the past, just as it is isolated today, was first discussed by Teichert (1958), who reached his conclusion on the basis of the tetrapods then known from Australia. Subsequent collecting strengthened Teichert's conclusion. For example, Anderson and Cruickshank (1978), who made an extensive study of Permian and Triassic tetrapods, found that Africa, India, and Eurasia were closely related then, just as they are today, but that Australia was essentially isolated.

Thulborn (1986) made a study of Australian Early Triassic tetrapods and found that the Australian fauna had a very different composition from that of coeval tetrapod faunas elsewhere. He found, first, that Australia's tetrapods are mainly amphibians, a fact that shows how different they are from their correlatives on other continents, where they are mainly reptiles. It is one of the very few places in Gondwanaland where the numbers of amphibians far exceed the numbers of reptiles. His second finding was that the Australian amphibian fauna, unlike that of other continents, is largely endemic at the family level. From these facts, Thulborn (1986) concluded that Australia was indeed isolated during Early Triassic time. Molnar (1992) came to the same conclusion. It is worth noting in passing that labyrinthodonts, which disappear from the record farther west by the end of the Triassic, persisted in Australia into the Early Cretaceous (Warren et al., 1991). In the Early Cretaceous in Australia, Molnar (1992) showed the continent to be largely isolated at that time.

In yet another study—this one an examination of India's paleopositions by Chatterjee and Hotton (1986)—all types of evidence were used, such as tetrapods, invertebrates, floras, stratigraphy, and magnetics. They concluded that India has always been close to Asia, even to Africa, but not to the other Gondwana continents. This is the same conclusion reached independent of, and using in part different data by, Gansser (1981), Waterhouse (1983), Stöcklin (1984), Saxena et al. (1986), and many others. (It should be noted that Chatterjee and Hotton's conclusions reinforce those by Anderson and Cruickshank.) If Chatterjee and Hotton (1986) are correct about India—and Gansser (written communication, 1987) feels very certain that his own similar conclusions are correct—then the significance of the magnetic anomalies of the Mid-Indian Ridge system must be reevaluated (Johnson et al., 1976; Meyerhoff and Meyerhoff, 1978; Agocs et al., 1992). The proximity of India to Asia also is shown by the similarity of their Late Cretaceous–Danian terrestrial vertebrate faunas (Gayet et al., 1984). Tetrapods seem to be giving the same message as invertebrates.

Floras

Smiley (1979, p. 311) wrote that "Floral distribution patterns for later Paleozoic and Mesozoic times conform with no continental drift model that has been proposed to date," a position he has convincingly documented in several outstanding papers (e.g., Smiley, 1974, 1979, 1992). In support of this statement, we mention the discovery of Late Permian *Glossopteris* and *Gangamopteris* floras, and even younger Gondwana floras (in addition to Gondwana marine faunas and tetrapods) in northeastern Asia—specifically, the Siberian platform (Tunguska area), Mongolian People's Republic, Kolyma River basin, Vladivostok area, Heilongjiang Province (China), and Korea (Neuburg, 1948; Zimina, 1967; Kon'no, 1968; Samylina and Yefimova, 1968; Meyen, 1969; Vozenin-Serra, 1984). Meyen (1969) personally checked and confirmed each of the Russian occurrences (Meyen, oral communication, 1974).

Meyen (1969), as well as Chaloner and Meyen (1973), warned that the identifications of *Glossopteris* and other Gondwana taxa in northern realms should be accepted with the greatest caution, and noted that they are most reliable where fructifications are present.

In further support of Smiley's (1979) statement, we mention a classic in-depth study of modern and fossil conifers and taxads by Florin (1963; see also Smiley, 1974, 1992). In the course of this study, Florin constructed many maps to show the northernmost and southernmost geographical limits of each taxon included in his study, for both living and fossil genera. The maps of the fossil and modern taxa are amazingly similar, a fact that suggests that the present latitudinal positions of land and sea might not have changed significantly since the Carboniferous, or at least that the distribution and dispersal mechanisms have remained unaltered to the present.

Finally, we mention the fact that many fossil floras are no more than assemblages of form "genera" and form "species" based on leaf and stem characteristics; they are not genera and species in the biological sense (Meyerhoff, 1952), despite a few opinions to the contrary (e.g., Kovacs-Endrody, 1991). As a consequence—and because many of the taxa that have been erected were named more than half a century ago—modern in-depth revisions of the taxonomy are much needed. Chandra and Surange (1979) have done a complete review of the Indian *Glossopteris* flora; Anderson and Anderson (1985) did a similar study of the coeval South African flora. A principal conclusion of both studies is that the taxa studied are endemic to the continent in which they were found, and are not conspecific with the taxa of other Gondwana areas. Without having many more fructifications than were studied, both by Chandra and Surange (1979) and by Anderson and Anderson (1985), such conclusions seem premature. Yet the fact that the authors of both monographs could not find any species similarities outside the area of each study is significant. We return to this topic in a subsequent section.

Bipolarity of global biotas and climate

A most important phenomenon that has been established by several detailed studies of fossil marine invertebrates, vertebrates, and floras is the bipolar distribution of many biotas from Late Devonian time to the present. A similar bipolarity ultimately may be established for older faunas and floras as well. This bipolar distribution of various biotas manifests itself as three distinct zones that are axisymmetrical about the *present* rotational pole, as was discussed in a preceding section. These include two polar biotal units, one Boreal and one Austral (Meridional), and one broad unit lying between the polar units—a warm-water or Tethyan-type unit. An obvious inference is that these zones directly reflect the climates of the times: cool climates at both polar regions and a warmer climate between. Data are as yet insufficient to quantify the temperatures of these various units for different times, except in a most approximate manner.

Milner's (1993) study of Upper Devonian tetrapods suggests a bipolar zonation for the Late Devonian. Hill's (1973) study of Lower Carboniferous corals suggests a bipolar zonation for Mississippian time. Smiley's (1979) study of fossil plants indicates bipolar zonations for Pennsylvanian and younger times until the present. Studies of Permian marine invertebrates and floras show distinct bipolar zonation for that period of time (e.g., Teichert, 1959, 1974; Florin, 1963; Dagis and Ustritskiy, 1973; Dutro and Saldukas, 1973; Ustritskiy, 1973; Smiley, 1974, 1979, 1992; Krassilov, 1987). Studies of marine invertebrates, vertebrates, and floras from Triassic time onward also show a distinct bipolarity (Florin, 1963; Kummel, 1969; Cariou, 1973; Dagis and Ustritskiy, 1973; Stevens, 1973; Ustritskiy, 1973; Dagis, 1974; Smiley, 1974, 1979, 1992; Crame, 1986; Krassilov, 1987; Yin Hongfu, 1990).

Conclusions

Runnegar (1979, p. 144) wrote that "Few geologists now seriously doubt that the Gondwanaland supercontinent existed in Late Palaeozoic to early Mesozoic time, and was fragmented by ocean-floor spreading in the Late Mesozoic and Cenozoic. The evidence is diverse, unambiguous, and decisive."

We quote this statement to make a point: the evidence summarized in this volume does not support Runnegar's assertion. Instead, we have concluded that, during all of Phanerozoic time, oceans were widespread in Pangaea, including Gondwanaland, as we have demonstrated earlier and in the pages that follow. How, then, did the tetrapods become so widespread? In our opinion, they were able to disperse just as the mammals have done throughout Cenozoic time—mostly by walking, but some by swimming, or even by rafting. Many reptiles—*Lystrosaurus* for example—have a skeletal morphology compatible with swimming. In fact, Colbert stated that ". . . *Lystrosaurus*, of Triassic age, appears to have been aquatic" (Colbert, 1969, p. 137). The times when land areas were high with respect to the oceans also should be noted (e.g., Late Permian–Early Jurassic; part of Eocene; Quaternary), for these are times when some tetrapods became relatively cosmopolitan. The example of the modern Galapagos Islands should be kept in mind, for this area has some 30 reptilian taxa which, as G. G. Simpson demonstrated (*in* Meyerhoff and Meyerhoff, 1974, p. 69), reached the Galapagos from Central America, more than 850 km away, by swimming. This is a much greater distance than that separating Antarctica from South America. Moreover, small tetrapods, particularly reptiles less than 1 m (3.3 ft) long, can and do travel long distances in the sea, with journeys lasting several months to two years (Leip, 1958). Leip (1958) reported the presence of the Tortugas loggerhead turtle, and other tropical sea turtles (all 46 cm [18 in] or shorter in length), from Cornish and Scottish beaches, having traveled distances of more than 9,500 km (5,890 mi)! None of the Antarctic Triassic tetrapods is as large as the largest Galapagos tetrapod (Kitching et al., 1972). With mammals there is less likelihood of such dispersal. One only needs to consider Madagascar with its highly endemic, almost Cretaceous-like lemur fauna (except for the pygmy hippopotamus whose ancestor was obviously an African Quaternary vagrant that swam over). Finally, we mention the strong possibility, based on new physical evidence, that a sizable isthmus joined Antarctica with South America before Late Jurassic time (I. Taner, oral communication, 1995).

In the sections that follow, we consider the individual regions.

FAUNAL AND FLORAL REALMS: FROM THE SUBCONTINENT TO NEW GUINEA

General

Current tectonic models interpret the Eurasian continent differently than the American and African continents. In Asia, a visible tectonic line, a suture, is drawn from southeastern Turkey to Papua New Guinea. Tethyan and northern realms, by definition, are north of this line; the Gondwanan and Malvinokaffric Realms are south. This tectonic line in south-central Asia was called originally the "*Indus suture*" by Gansser (1964, p. 62). We call it the "*Indus-Yarlung suture zone*." (The commonly used expression Indus-Tsangpo suture is etymologically wrong, because Tsangpo is a Tibetan word for river; see Meyerhoff et al., 1987. The river called the Tsangpo is the Yarlung River). West of Pakistan it is called the "*Zagros suture zone*." East of India and Bangladesh, the suture has no general name. The suture zone is marked by scattered and discontinuous outcrops of ophiolites. In plate tectonics, this is the zone along which Eurasia and the landmasses south of it (Afro-Arabia, India, Australia) collided in Cretaceous or early Tertiary time, being previously separated from Eurasia by a deep Tethys ocean (Drewry et al., 1974; Johnson et al., 1976).

No equivalent line exists along the suture's extensions in either Africa or the Americas. Instead, the feature separating northern from southern biotic realms is a vague, indefinite zone cutting across Sahara sands in Africa and dense jungle in South America.

An outstanding example of this overlapping or intercalary relationship is the lowermost Permian cold-water sequence found in Arabia, Pakistan, Peninsular and Himalayan India, and in Tibet (Dickins and Shah, 1979, 1981, Dickins, 1985b, Kapoor and Tokuoka, 1985; Dickins et al., 1993). Where it was first studied in northern Pakistan and India, the pebbly mudstone unit (at the bottom of the mudstone-carbonate sequence) grades from a continental tillite deposit to a neritic shelf deposit containing erratic pebbles and boulders (Teichert, 1967). Although pebbly mudstone is widespread, it need not everywhere be a glacigene deposit (Waterhouse, 1982). Its fauna is of Gondwana aspect, and has a low taxic diversity. In contrast, the carbonate-bearing unit above was deposited in shallow, warm-water, tropical-subtropical seas. Its fauna is rich, diversified, and Tethyan (Teichert, 1967, 1981; Waterhouse, 1982; Dickins, 1985b).

The pebbly mudstone and overlying carbonate-bearing unit occupy a band several hundred kilometers wide (in a north-south direction). The widest part is in northern India, Kashmir, Pakistan, and Xizang (Tibet), where the two units have been traced across a zone about 1,000 km (620 mi) wide. The biotas associated with the two units have an even wider spread, reaching nearly 3,500 km (2,170 mi) between Umaria, Madhya Pradesh (atop the Indian shield; Fig. 9), and Urumqi, northern Xinjiang Uygur Autonomous Region (Fig. 1). Many articles have been written on these Early Permian units (Hudson and Sudbury, 1959; Hudson, 1960; Gansser, 1964; Teichert, 1967, 1981; Kummel and Teichert, 1970b; Shah and Sastry, 1975; Grant, 1976; Acharyya et al., 1979; Dickins and Shah, 1979, 1981; Nakazawa and Kapoor, 1979; El-Khayal et al., 1980; Cheng Zhengwu, 1981, Jin Yugan, 1981, 1985; McClure and Young, 1981; Page, 1981; Srikantia, 1981; Stauffer and Mantajit, 1981; Chen Bingwei, 1982; Waterhouse, 1982; Liang Dingye et al., 1983; Liu Benpei and Cui Xinsheng, 1983; Wang Yujng and Mu Xinan, 1983; Yin Jixiang et al., 1983; Fontaine and Vachard, 1984; Hu Changming, 1984; Ingavat, 1984; Metcalfe, 1984; Vozenin-Serra, 1984; Zhao Junpu, 1984; Matsuda, 1985; Nakazawa and Dickins, 1985; Pakistani-Japanese Research Group, 1985; Huang Jiqing and Chen Bingwei, 1987; Fang Zhongjie, 1993). The pebbly mudstone unit and its equivalents are underlain by early Paleozoic carbonates with northern faunas.

Although the intercalary zone containing the Early Permian cold intercalation is the most studied of this vast region, we mention that post–Early Permian warm-water Phanerozoic sections are the rule here; the Gondwana cold intercalation is a marked exception. This entire intercalary zone has been called the northern shore of Gondwanaland, or the southern shore of Tethys. We have never seen it called the southern shore of Angaraland (or of Laurasia), although this makes just as much sense. Some have even compared it with a faunal province or realm (e.g., Perigondwana province, Matsuda, 1985; Okimura et al., 1985; Peri-Gondwana Tethys: Nakamura et al., 1985). Archbold (1983) noted that it coincides partly with the Cimmerian province of some literature.

One phenomenon related to the intercalary zone has always intrigued us. When a representative Tethyan taxon is found in, say, a New Zealand setting, it becomes evidence of a strange terrane, an exotic terrane, or a tectonostratigraphic terrane (e.g., Spörli and Gregory, 1981). When the opposite situation is encountered—i.e., a Gondwana taxon is found in northern Xizang (Tibet)—the whole suture zone between the southern and northern continents is shifted northward for whatever period of time is necessary. This practice makes no sense to us.

Regardless of what has been done in the past, it is obvious that a new approach is in order. Biogeographical data are powerful tools for resolving various geological and geophysical problems. Whatever the intercalary zone between Tethys and Gondwana may be, it is not a suture zone of the type envisioned in the paleogeographies of Dietz and Holden (1970), Drewry et al. (1974), and Johnson et al. (1976).

We have commented on the fact that India and Asia have never been separated by any great distance (Meyerhoff and Meyerhoff, 1978; Haller, 1979; Auden, 1981; Dickins and Shah, 1981; Gansser, 1981, 1991; Waterhouse, 1983; Stöcklin, 1984; Chatterjee and Hotton, 1986; Saxena et al., 1986; Zheng Haixiang and Zhang Xuanyang, 1990). Gansser (1981, p. 111) wrote that "the known structural and stratigraphic facts require that Peninsular India was never far distant from a most complex southern front of Eurasia," and has since (Gansser, 1991, p. 47) presented additional compelling evidence to support his views. Stöcklin (1984, p. 65), writing an outstanding summary of the field data that show India always to have been close to Asia, stated that the plate tectonic concept of India ". . . is in flagrant contradiction with the total lack of geological evidence for the existence of a Permo-Scythian Tethys Ocean and for any substantial development of an oceanic Neotethys . . ." Auden (1981), in a comprehensive review of all of the circum-Indian geology and geophysics, concluded (p. 628) that ". . . India has had a relatively close association with Eurasia throughout the Phanerozoic, notwithstanding the allowance which must be made for crustal shortening during the formation of the Himalayas." Even A. B. Smith (1988), using paleontological data, showed that plate tectonic models requiring the presence of a broad, deep ocean between India and Asia are untenable because of the gradual, not abrupt, faunal changes that take place across the Qinghai-Xizang (Tibet) Plateau and the Himalayas. Similarly, in the border region between southwestern China and Burma, and between southwestern China and Thailand, Liu Benpei et al. (1991) have demonstrated the presence of a broad transitional band separating Gondwana faunas and floras on the southwest and west from Tethyan and Cathaysian faunas and floras in the north and east. Despite geological and paleontological field data, however, the Drewry et al. (1974) model and its variants are still used in this area where the facts can no longer be explained by former wide separations (e.g., not more than 500 km [310 mi]; Gansser, 1981, 1991).

Of all the data that show India's proximity to Asia, the pre–Early Permian stratigraphy and biota are among the most important. Upper Proterozoic through Mississippian strata of the *same formations and faunal realms* are widespread on both sides of the Indus-Yarlung suture and its western and eastern extensions to Turkey and Papua New Guinea, respectively (Voskresenskiy et al., 1971; Waterhouse, 1979, 1983; Jin Yugan, 1981, 1985; Chen Tingen, 1983; Yin Jixiang et al., 1983; Bhatia, 1984; Vozenin-Serra, 1984; Mu Enzhi et al., 1986; Saxena et al., 1986, Kumar et al., 1987; Smith, 1988; Liu Benpei et al., 1991; Meyerhoff et al., 1991). Further, there is a unique interfingering of biota from different biogeographical realms within these strata (see references above).

The most common partial explanation of these facts postulates that the Indian subcontinent was attached to Gondwanaland until Early Permian time (Huang Jiqing, 1984; Smith, 1988); this explanation, however, ignores the pre-Permian problems. According to this idea, Gondwanaland drifted rapidly northward, attaching itself to Asia along the Litian–Jinsha Jiang (Longmu Co-Yushu) suture zone in northern Xizang (Tibet) and north of the Indus-Yarlung suture. The collision between Gondwanaland and Asia would have taken place during Late Permian time. Then the Indus-Yarlung suture opened during Late Triassic time, and India moved south, only to shuttle northward again to close the suture zone in middle to late Eocene time.

Taking into account the geology of the region, Mu Enzhi et al. (1986, p. 34) wrote that, "Probably the most difficult aspect of the Himalayan Paleozoic to explain has been the intimate interbedding of Gondwanaland glacial-marine beds and cold-water faunas. It makes little sense, of course, to move the Himalayan region rapidly into and out of the Gondwanaland cold region in a purely mechanical sense because of the great distances and short time involved for the to-and-fro movements. But it is reasonable to try to explain the major climatic change in terms of an oceanic surface current alteration of major proportions that was able to shunt a cold-water current into a warm-water, low-latitude region and maintain it for at least a few million years. Such an hypothesis would explain the interbedding of cold and warm phenomena in the Himalayas. . . ."

Indian subcontinent–western China

Pre-Carboniferous. Peninsular India lacks any fossiliferous marine beds of pre-Permian, Phanerozoic age (Figs. 4 through 7a,b). Western China and adjacent regions, however, are rich in fossiliferous northern region beds of Cambrian through Devonian age (Figs. 4-7a,b). Faunas from various regions of the Old World Realm dominate the whole of Asia north of India (e.g., Boucot et al., 1969; Boucot and Gray, 1979; Boucot, 1985, 1988; Hou Hongfei and Boucot, 1990) (Figs. 7a,b).

Carboniferous. During Mississippian time, the sea occupied much of western China, Nepal, northernmost India and Pakistan, and Kashmir. The marine invertebrate faunas are rich in brachiopods, conodonts, crinoids, bryozoans, ostracodes, corals, and algae of warm-water Tethys affinities (Mamet and Belford, 1968; Waterhouse, 1979; Srikantia, 1981; Liang Dingye et al., 1983; Yin Jixiang et al., 1983). The corresponding land flora also was a cosmopolitan type (Pal and Chaloner, 1979; Zhang Shanzhen, and He Yuanliang, 1985).

In Pennsylvanian (Bashkirian and younger; Fig. 3) time, differentiation into distinct realms began. The marine invertebrate faunas of northernmost Pakistan and India, Kashmir, Nepal, and Xizang (Tibet) are closely related to, and in some cases are congeneric and conspecific with, taxa in southern Europe, the Russian platform, the American Midcontinent, and, most recently, Cuba (Mu Enzhi et al., 1973; Waterhouse, 1979; Yin Jixiang et al., 1983; Bhatia, 1984; Pszczolkowski, 1989).

Recent stratigraphical and paleontological studies in the mountainous region along the Yunnan (southwestern China)-Burma-Thailand border region have revealed the presence there, in a north-south band several hundred kilometers wide, of extensive interlayering of southern and northern faunas and floras in the Viséan-Asselian section (Figs. 1, 3) (Han Neiren et al., 1991; Liu Benpei et al., 1991). As much as 28% of the palynoflora in southwestern Yunnan is Gondwanan, with the percentage of Cathaysian and Tethyan taxa increasing north- and eastward, and the percentage of Gondwanan taxa increasing west- and southwestward (Liu Benpei et al., 1991). The faunas also grade north- and eastward to Tethyan and westward to Gondwanan. Carboniferous cosmopolitan taxa cross from Gondwanan to northern areas with little change in composition (Han Neiren et al., 1991; Liu Benpei et al., 1991).

Permian. During Early Permian time, many parts of the Indian subcontinent underwent mountain glaciation (Figs. 8 through 10) (Meyerhoff and Teichert, 1971). The diamictites that formed during this glaciation are known by a variety of names: the Tobra Formation in northern Pakistan (Kummel and Teichert, 1970a,b); the Talchir boulder beds in the Jodhpur area

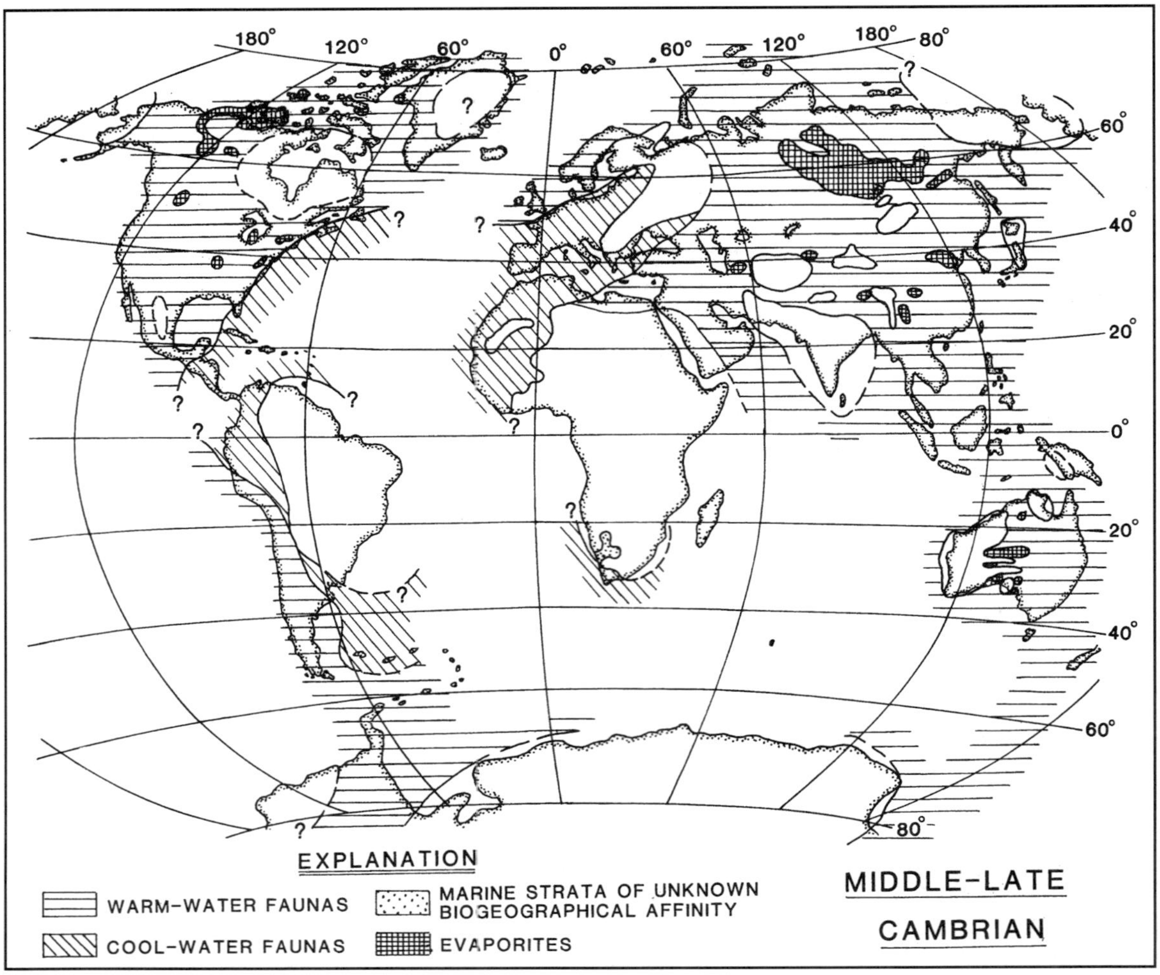

Figure 4. Paleobiogeographical units of Middle-Late Cambrian time. Based in part on Boucot and Gray (1983).

of Rajasthan, western India (Fig. 9); the Bap Formation in southern Rajasthan (Ranga Rao et al., 1979); and several other local names. The Talchir perhaps is the most common name. (The Himalayan Blaini boulder beds, once thought to be equivalent to the Tobra and Talchir, are now known to be of late Proterozoic to earliest Cambrian age [Brasier and Singh, 1987; Brookfield, 1987].) The faunas and floras associated with the glacial and glacigene strata, including the overlying *Conularia-Eurydesma* beds, have typically Gondwana-type taxa, not just in Kashmir, but in all other parts of India and Pakistan where beds of this age are present (Teichert, 1966, 1967, 1981; Kummel and Teichert, 1970b; Nakazawa and Kapoor, 1979; Ranga Rao et al., 1979; Sohn and Chatterjee, 1979; Srikantia, 1981; Bhatia, 1984; Kapoor and Tokuoka, 1985; Nakazawa, 1985; Li Xingxue and Wu Xiuyuan, 1993).

The *Conularia-Eurydesma* beds are overlain by a succession of strata of early Artinskian, Late Permian, and Triassic ages (Figs. 3, 10). Throughout the northern India–Kashmir–Salt Range province, the overlying Lower and Upper Permian sections have major carbonate units rich in tropical-subtropical Tethys faunas (Fig. 9). The same general sequence has been described toward the east, as far as Arunachal Pradesh and Burma at the northeastern corner of India (Fig. 9) (Singh, 1978, 1979; Acharyya et al., 1979; Waterhouse, 1979). However, the carbonates of the northern Pakistan–Kashmir area are not as abundant and are replaced laterally in places by mafic volcanic rocks. Northwest of the Salt Range area, Permian Gondwanan and Tethyan faunas also are present in the Pamir Range of Tajikistan (Fig. 9) (Grunt and Dmitriyev, 1973). The *Eurydesma* fauna reported by Liu and Cui (1983) in the Rutog area of western Xizang (Tibet) has since been claimed by Fang Zhongjie (*in* Liu Benpei et al., 1991, p. 26) to be an identification error, although Dickins disagrees with Fang and considers the material to belong to *Eurydesma* (see also Li and Wu, 1993).

A most interesting set of coeval strata, mainly equivalent to the Tobra and overlying Lower Permian beds, has been found with marine tongues south of the Salt Range in northern Pakistan to the vicinities of Manendragarh and Umaria atop the Indian Shield, 1,200 km (745 mi) south-southeast of the Salt Range (Fig. 9). Here, in central Madhya Pradesh, Asselian to Sakmarian

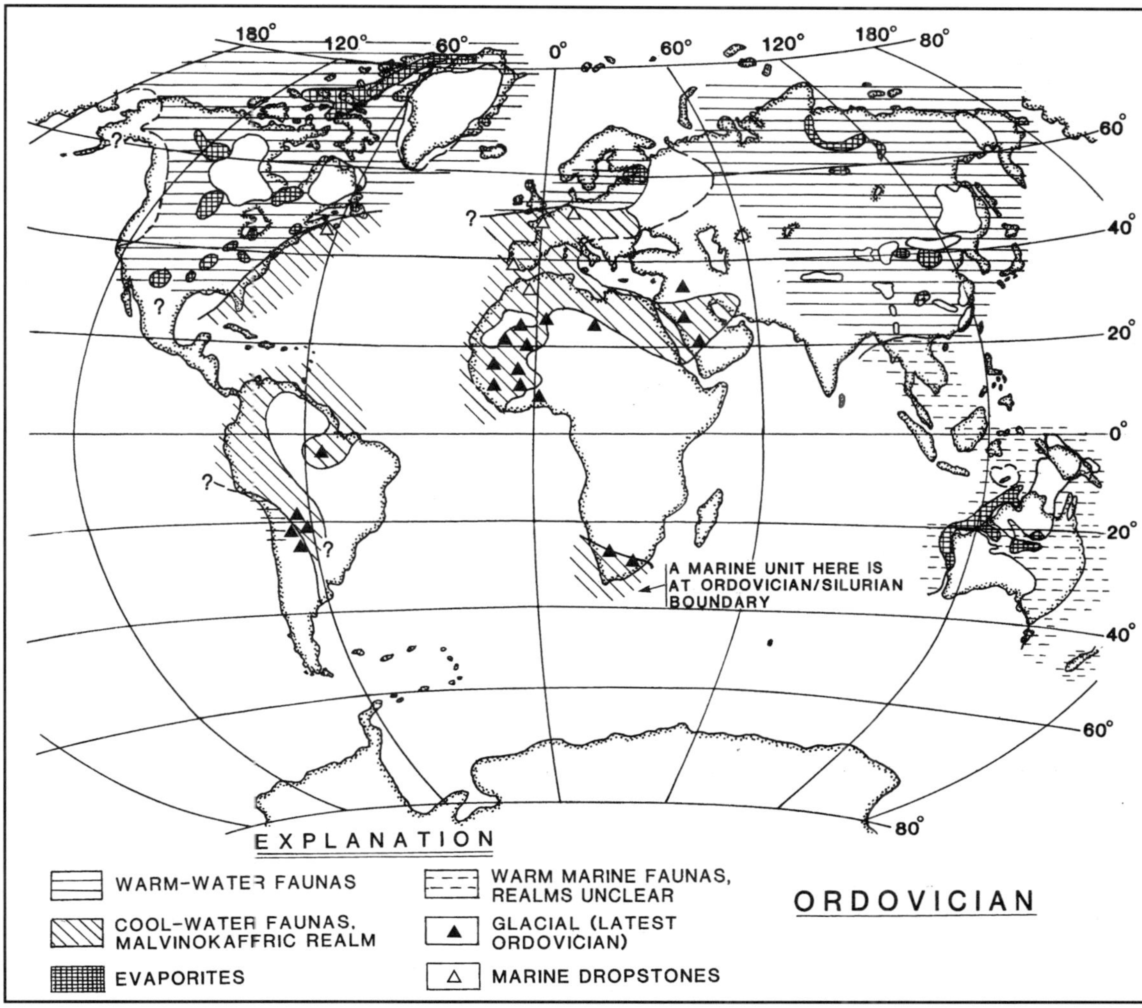

Figure 5. Paleobiogeographical units of Ordovician time. Based in part on Boucot and Gray (1983).

(Fig. 3) marine strata at Manendragarh are interbedded with the Talchir boulder beds. Elements of the Tobra fauna have been described from Victoria, southeastern Australia (Archbold, 1991a). The Permian marine tongues in Peninsular India apparently came from the ocean east of the subcontinent (Shah and Sastry, 1975). A short distance farther west at Umaria in the upper Narmada Valley, a younger Sakmarian marine intercalation is present (Reed, 1928) (Figs. 3, 9). This sea came from the west of India and is believed to have connected to the Salt Range region via equivalent strata with marine tongues in southwestern Rajasthan (the Bap Formation at Bhadaura) and in the Karampur (Karmpur) well, drilled as an exploratory test for petroleum south of the Salt Range (Fig. 9) (Voskresenskiy et al., 1971; Shah and Sastry, 1975; Dickins and Shah, 1979; Ranga Rao et al., 1979; Dickins, 1985b). Another marine tongue has been found in southern India in the Palar basin around Madras (Fig. 9). This tongue, like that at Manendragarh, transgressed from the east (Murthy and Ahmad, 1979), as did still another tongue in Bihar in northeastern India (Fig. 9) (Dutt and Shah, 1969).

North of the subcontinent, Early Permian pebbly mudstones with Gondwana fauna and flora and later Permian strata (Fig. 10) with a richly diverse Tethys fauna have been found at scattered localities almost into the Kunlun Shan along the southern rim of the Tarim basin, 600 to 900 km (370 to 160 mi) north of the Salt Range (Figs. 1, 9) (He Xinye and Weng Fa, 1983; Liang Dingye et al., 1983; Liu Benpei and Cui Xinsheng, 1983; Wang Hongzhen, 1983; Wang Yujing and Mu Xinan, 1983; Jin Yugan, 1985; Hsu Ren et al., 1990). In fact, wherever cold-water Gondwana faunas of the Early Permian are present, they are succeeded abruptly by Tethyan warm-water faunas—from the subcontinent to New Guinea and Western Australia, as well as from the subcontinent to the Andes (Fig. 10).

In the Kuruk Tagh (Fig. 1), an offshoot of the Tian Shan in the northeastern part of the Tarim basin, Norin (1930) found a diamictite (which he claimed is a true tillite) overlying Early Carboniferous, slightly metamorphosed, coal-bearing strata and underlying fossiliferous marine Permian strata. D. Hill (1958) wrote that the Kuruk Tagh glacial deposits are present in a fair-sized area between 88° and 90°E and 40° to 42°N. She concluded, on the basis of stratigraphic position and regional

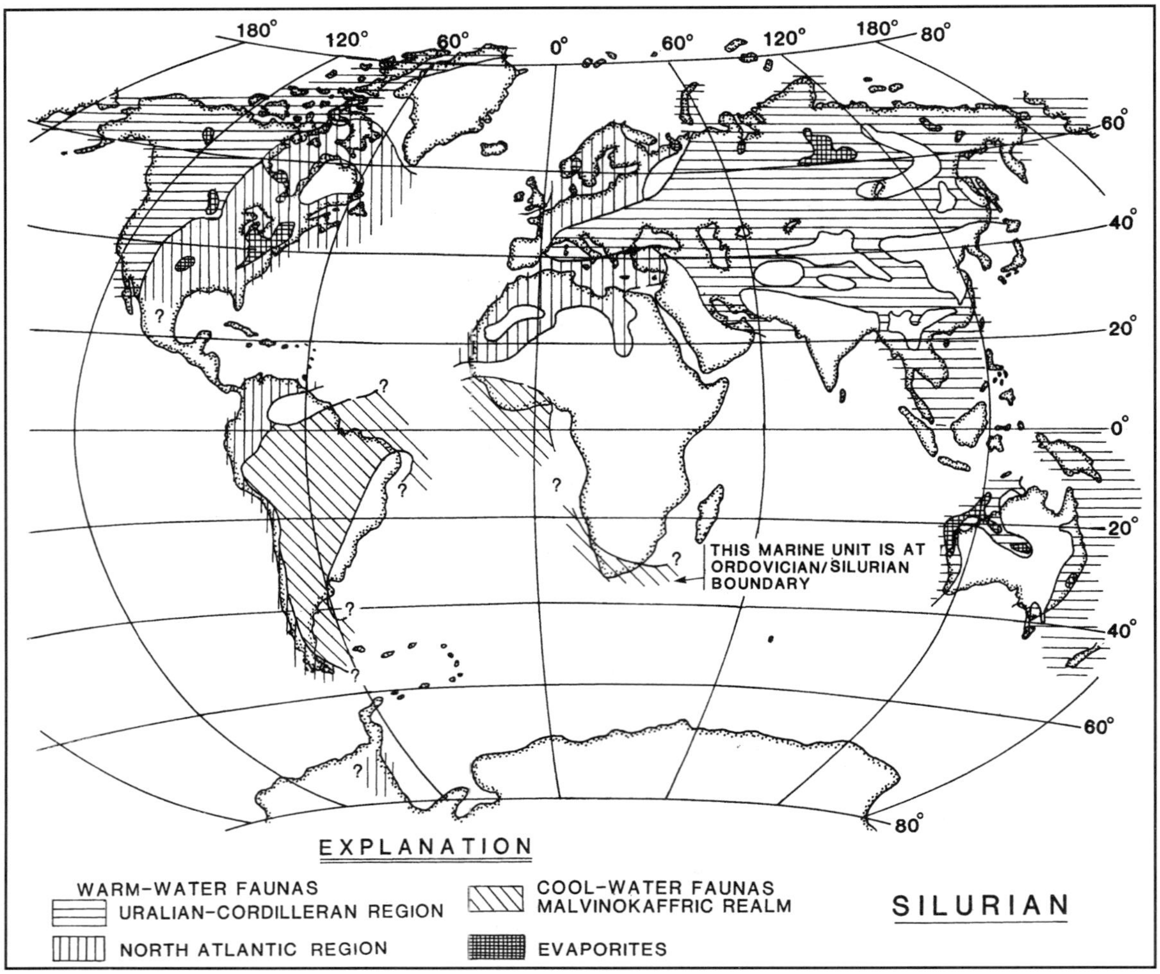

Figure 6. Paleobiogeographical units of Silurian time. Based in part on Boucot and Gray (1983).

paleogeography, that the diamictites are of Sakmarian age (an Asselian age is not precluded).

Ustritskiy (1973) observed that faunal distribution during all of Permian time was bipolar, just as today. He reported that Permian marine invertebrate faunas, Austral and Boreal, are cool-water forms, also as today. The faunas are of low taxic diversity; large numbers of Tethyan taxa are absent; cool-water agglutinated foraminifers are the rule; fusulinids, except in Novaya Zemlaya, as well as tabulate and rugose corals, are absent. The same bipolar distribution continued through both the Mesozoic and Cenozoic (Florin, 1963; Ustritskiy, 1973; Dagis, 1974; Krassilov, 1987; Yin Hongfu, 1990).

The plants show generally the same picture as the marine invertebrates, and exhibit a high generic and specific diversity (Bose et al., 1989). In the Salt Range, the *Glossopteris* flora is succeeded by an Angaran flora (Smiley, 1979). Some strata contain admixtures of all three floras of Permian Asia—Gondwana, Angara, and Cathaysia (Lacey, 1975; Kapoor, 1979)—a situation that is repeated across a belt nearly 3,000 km (1,860 mi) broad from northern Xinjiang Uygur Autonomous Region to Peninsular India (Maithy, 1976; Zhang Lujin, 1983a). Many localities containing Gondwana floras (including *Glossopteris*) have been found in Xizang (Xu Ren, 1973, 1976; Chen Bingwei, 1982; Yin Jixiang et al., 1983; Hu Changming, 1984; Zhao Junpu, 1984; Zhang Shanzhen and He Yuanliang, 1985; Hsu Ren et al., 1990; Li Xingxue and WuXiuyuan, 1993). Many of the taxa found in Xinjiang near the towns of Urumqi and Karamay are conspecific with Indian taxa (Zhang Lujin, 1983a, 1990). This last fact indicates the contiguity of India with southern Asia (Chandra and Surange, 1979).

The presence of *Glossopteris* in places with *Gangamopteris* has now been proved in Late Permian deposits of Korea, eastern Manchuria, and the neighboring Primor'ye region of the USSR, just north of Vladivostok, 4,000 to 4,400 km (2,500 to 2,700 mi) northeast of India (Neuburg, 1948; Zimina, 1967; Kon'no, 1968; Meyen, 1969; Vozenin-Serra, 1984). Similarly, *Glossopteris* has been identified in the Podkammenaya Tunguska-Tunguska Rivers area of the Siberian platform, and in the Mongolian People's Republic (Fig. 1) (Zimina, 1967; Meyen, 1969).

The vertebrate fauna tells much the same story—a very close relationship with Europe and Africa, as well as with North America (Table 1). At least one dicynodont species, *Diictodon tienshanensis*, is common to China and South Africa (Cluver and Hotton, 1979). Chatterjee and Hotton (1986) concluded that India always has been physically very close to Asia, not to Australia and Antarctica, and the vertebrate data fully support their conclusions. Kutty's (1972) study of Permian tetrapods and the investigations of Gayet et al. (1984) latest Cretaceous-Paleocene taxa led to an identical conclusion.

Triassic. Beginning in Late Permian time, the seas began a general withdrawal from much (but not most) of what is now China, and continental conditions set in over a huge area (Figs. 11, 12a,b). Equally continental conditions persisted on the Indian Peninsula. However, between the two continental areas, Triassic marine Tethys facies is dominant and widespread. One of the thickest and least studied marine Triassic sections on Earth crops out across huge areas in the mountains of southwestern China (Meyerhoff et al., 1991). The marine faunas are rich and varied, and show that reproductive communication between Europe, Southeast Asia, and Western Australia was almost uninterrupted (Kummel, 1969; Kummel and Teichert, 1970a,b; Nakazawa et al., 1975; Nakazawa and Kapoor, 1979, 1981; Waterhouse, 1979; Srikantia, 1981; Bhatia, 1984; Dickins, 1985c; Kapoor and Tokuoka, 1985; Nakazawa, 1985). Kummel's (1969) classic monograph on Scythian ammonoids (Fig. 3), one of the few studies of its kind in the world, demonstrates clearly the close links among Germany, the Balkans, Greece, the Russian platform, the Caspian Sea region, Afghanistan, northern Pakistan, Kashmir, and Southeast Asia (Fig. 1). Ostracode studies show only close connections with northern continents, not with the southern continents (Sohn and Chatterjee, 1979; Bhatia, 1984). Dagis (1974) recognized Austral and Boreal Triassic units additional to Tethyan. These Austral and Boreal units have the same bipolar positions as today.

Plant studies are at least as definitive. Zhang Lujin (1983b) found a Triassic microflora in the Urumqi area, Junggar basin (Fig. 1), northwestern China, which is a mixture of Cathaysian, Gondwanan, and Angaran elements; in addition it has a strong Indo-European component. For example, of the 81 Jurassic genera from India discussed by Smiley (1979), 48 are Euramerian (Euramerican) and 29 are Gondwanan. Of the 29 Gondwana genera, only 5 are limited elsewhere to Gondwana landmasses; the others occur in transitional and northern areas as well. This is why Russian paleobotanists traditionally treat India as a part of the European province, calling it instead the Indoeuropean Province of the Euramerian Realm. This realm, according to Meyen (1969), existed from Pennsylvanian time onward, before which time the plants of the world were more cosmopolitan. Rhaetian (Fig. 3) microfloras on the Lhasa block of Xizang are related to microfloras in Canada, Greenland, Western Europe, the Caucasus, and Southeast Asia (Adloff et al., 1984). The Upper Triassic microflora of Kutch (Kachchh) on India's northwestern coast is similar to that of Western Europe, Arctic Canada, and the Salt Range, not that of Gondwana proper (Koshal, 1984). Gondwana floras on the Indian craton share many European taxa.

The vertebrates show free reproductive communication with Europe, Africa, eastern Asia, and North America (Table 1) (Chatterjee and Hotton, 1986). One well-known tetrapod, the Early Triassic *Lystrosaurus*, is present in India, Africa, Antarctica, Vietnam, western China, and the Moscow basin. The occurrences in the former USSR were described by Kalandadze (1975). The Chinese section has many of the standard South African vertebrate zones, *Lystrosaurus* being the oldest, probably ranging into the youngest part of the Permian. Kannemeyerids from the African *Cynognathus* zone of the overlying Middle Triassic are numerous (Yuan and Young, 1934; Sun Ailin, 1973a,b; Kalandadze, 1975; Zhou Mingzhen, 1981; Kalandadze and Rautian, 1983; Chatterjee and Hotton, 1986). *Lydekkerina*, an Early Triassic genus of the *Lystrosaurus* zone, is known from India, South Africa, and Antarctica. The kannemeyeriids are known from China, South and East Africa, the European Russia, and North and South America. Several other representatives of the *Cynognathus* zone are present in China, including a procolophinid, a theriodont that is very close to *Cynognathus*, and an erythrosuchid. Upper Triassic Indian forms also have representatives in China. More recently, Chatterjee (1986a,b) found Indian Late Triassic *Malerisaurus*, and three more taxa, in Texas, south-central United States (Fig. 2). *Malerisaurus* now has been found on all continents except South America. The Indian tetrapods are typical of Eurasia, coming close to being globally cosmopolitan; hence it is unreasonable to conceive of India having been detached from Asia and the other continents. In another example, Tarlo (1959, 1967) found typically Tethyan nothosaurs from the Middle Triassic of northeastern India. This close relationship of India with the north is repeated in every part of the fossil-bearing section, leading Chatterjee and Hotton (1986, p. 165) to conclude that "India was a cross-road of faunal migration between Africa, Asia . . .," just as it is today, which fits very well with the overall cosmopolitaneity of the Triassic terrestrial vertebrates.

Jurassic-Cretaceous. The neritic to shallow-water Cretaceous of northern Pakistan, northern India, Kashmir, and the Himalaya, as well as of the Lhasa block north of the Indus-Yarlung suture zone, belongs to the shallow, tropical, warm-water Tethys (Figs. 1, 12a,b, 13). The *Orbitolina*-bearing and rudistid carbonates are typical of the same facies in the Mediterranean and the Caribbean (Mu Enzhi et al., 1973; Srikantia, 1981; Jaeger et al., 1982; Yin Jixiang et al., 1983; Bassoullet et al., 1984; Pons and Vozenin-Serra, 1984; Pudsey et al., 1986). Deep-water deposits recently described from Xizang (Tibet) contain well-preserved Upper Jurassic radiolarians with distinctive Tethyan affinities (Yang Qun and Wang Yu-jing, 1990).

The Tethys facies extends down the western and eastern sides of the Indian shield. In the west in Kutch (Kachchh), Spath (1928) described 556 ammonoid species, of which 122 (22%) are conspecific with ammonoids in Europe (Neaverson, 1955; Arkell, 1956; Mehra et al., 1979). Open-sea connections also extended to Madagascar, the Far East, and the Himalaya Tethys (Arkell, 1956; Shah and Sastry, 1981).

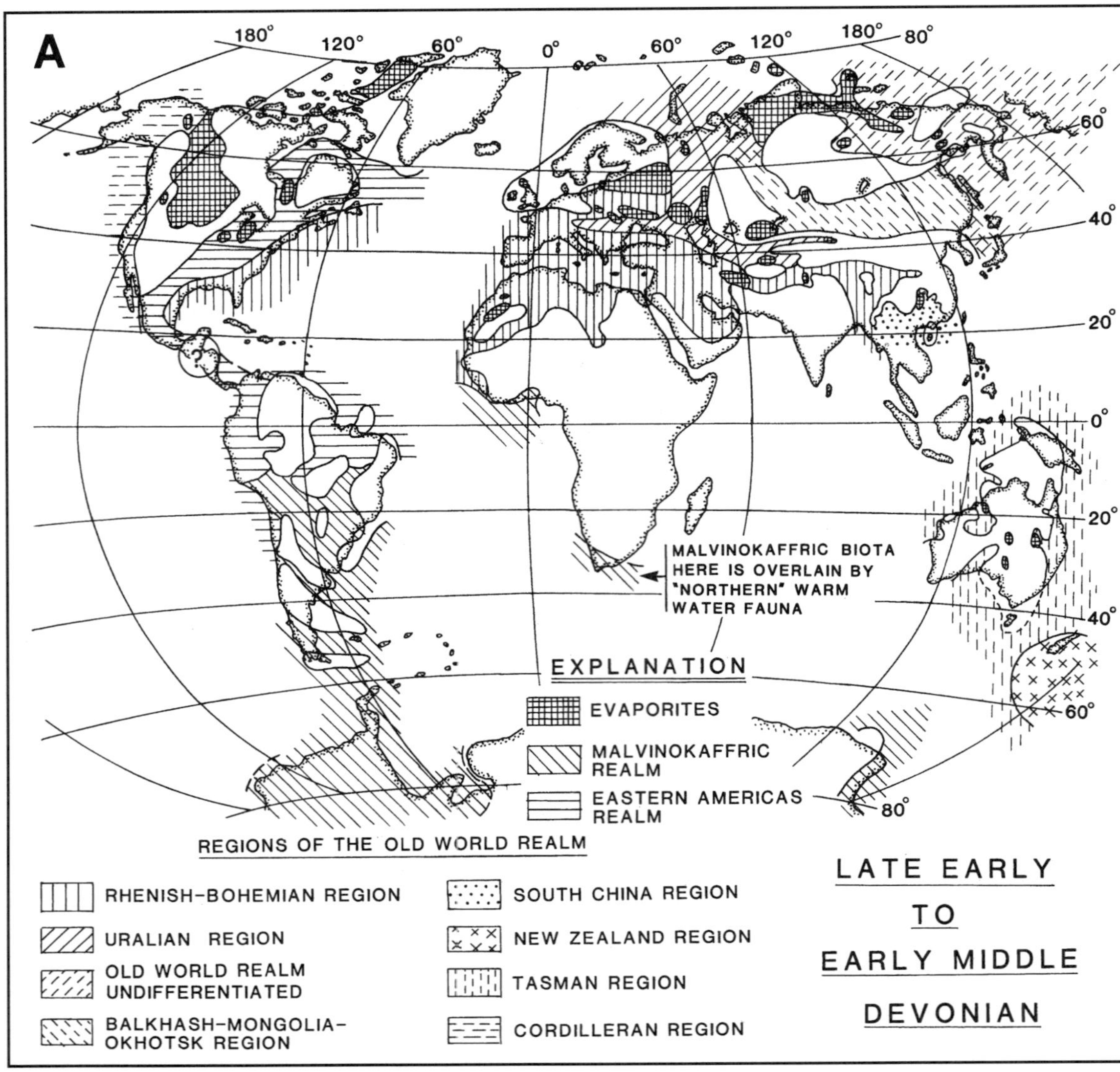

Figure 7. A, Paleobiogeographical units of late Early to early Middle Devonian (Emsian-Eifelian) time. Based in part on Boucot and Gray (1983). B (on facing page), Paleogeographical map modified from Drewry et al. (1974) showing late Early to early Middle Devonian biogeography and evaporite distributions. In our view, the data of Drewry et al. (1974) are flawed insofar as there is no convincing evidence for the presence of high-latitude South American Devonian tillites, nor for the presence of oceanic opaline silica occurrences separate from continental margin lydites. Furthermore, the lydites are interbedded in many places with clearly continental siliciclastic rocks. The reconstruction of Drewry et al. is incompatible in terms of rational oceanic circulation patterns with the known biogeography of the later Early Devonian–earlier Middle Devonian, particularly in regard to the varied regions of the Old World Realm and the relations of the Old World and Eastern Americas Realms.

Several Mesozoic-Tertiary basins occupy India's eastern coast from Bengal in the north to Sri Lanka in the south. All have Tethyan Jurassic and Cretaceous marine facies (Raju, 1968; Ramanthan, 1968; Sastri and Raiverman, 1968; Sastri et al., 1977; Rasheed, 1978; Murthy and Ahmad, 1979; Chiplonkar and Phansalkar, 1980; Bhattacharya, 1981; Kumar, 1983). Many ammonoid, foraminiferid, and other taxa are congeneric with taxa in Europe, the Mediterranean, the Caribbean, Mexico and the U.S. Gulf Coast (Rasheed, 1978; Chiplonkar and Phansalkar, 1980), although generic similarities tend to decrease with decreasing time because of increasing levels of provincialism toward the end of the Mesozoic. Small reefs and evaporite lagoons are found around the present coast, a fact that suggests both elevated temperatures and close proximity of the ocean. Early and Middle Jurassic (Tethyan) waters penetrated up to 320 km (200 mi) into the present peninsula near Sirpur (Fig. 9) (Bhattacharya, 1981). Yet this is a part of Gondwana that was supposed to have been about 6,000 km (3,700 mi) south of its

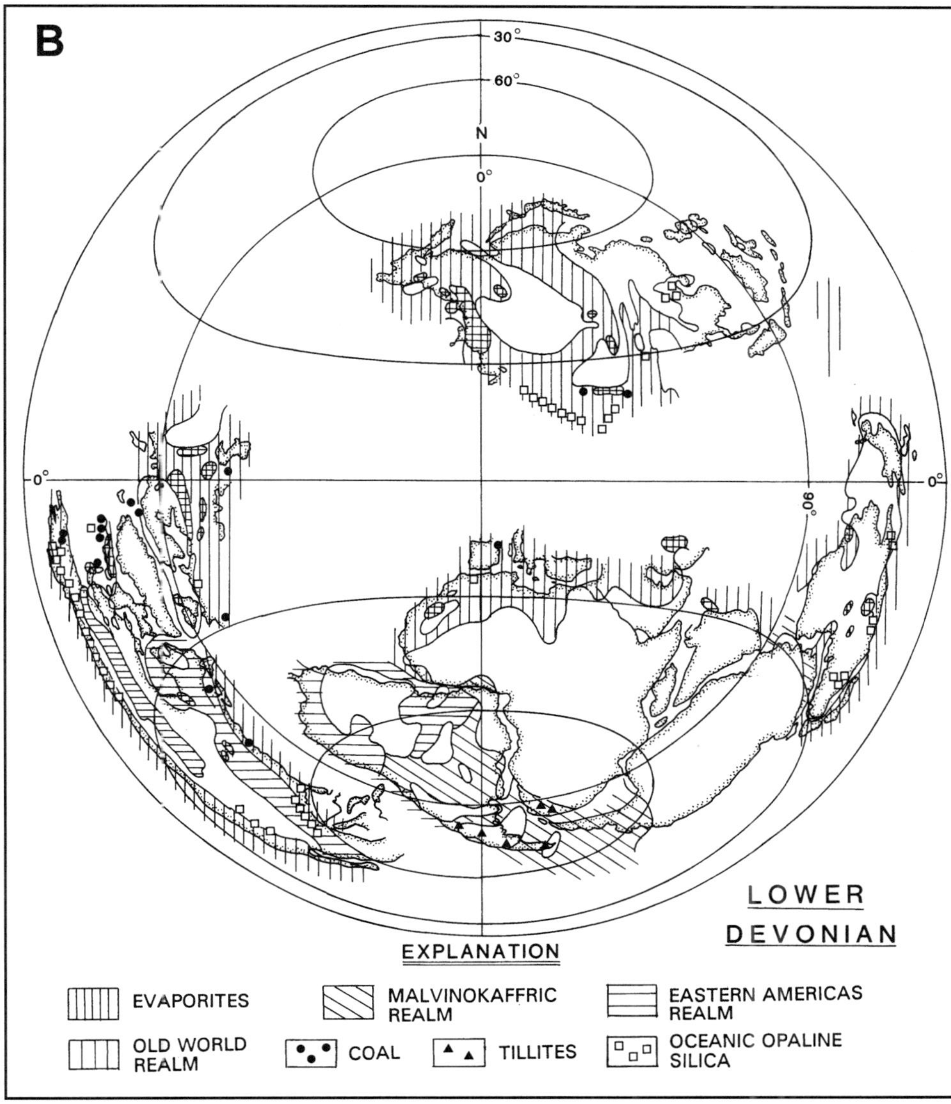

present position and far from any part of the world ocean (Johnson et al., 1976, p. 1562)!

The freshwater faunas (ostracodes, fish, amphibians, reptiles) of the Indian shield show almost no endemism, and nearly all taxa studied have their closest affinities in northern realms (Govindan, 1975; Jain, 1980, 1983; Bhatia, 1984; Schaeffer and Patterson, 1984; Patterson and Owen, 1991). The marine ostracodes are related to species and genera in China and Europe, with one taxon known also in Ethiopia and another in South America (Bhatia and Rana, 1984). Some taxa are known from Mongolia and the Russian platform (Govindan, 1975). Even the Late Cretaceous freshwater taxa, although with links to peripheral areas of Gondwana, show strongest links to the northern continents (Gayet et al., 1984); many genera, even species, from India are known in Mongolia (Starkevich, 1982).

In northeastern China, in the Wanda Shan of eastern Heilongjiang Province, north of Vladivostok, Gu Zhiwei et al. (1984) found Gondwana species of pelecypods in Jurassic marine tongues, demonstrating that this northeastern part of Asia maintained links with Gondwana, as it did during the Permian and Triassic (Fig. 1).

The fossil plants show a similar picture. They have Gondwana elements, but wherever they are found, they are dominated by genera common in central and northern Xizang, Europe, North America, China, Japan, and Siberia, with only limited numbers from India (Tuan Shuyin et al., 1977; Smiley, 1979; Pantic et al., 1981; Gansser, 1983; Pons and Vozenin-Serra, 1984; Hsu Ren et al., 1990).

Far to the northeast in the Kolyma drainage basin of northeastern Siberia, north of Magadan, Samylina and Yefimova (1968) have described a Liassic Gondwana flora of *Dicroidium*, *Ptilophyllum*, and other genera (Fig. 1).

As for the vertebrates, Table 1 tells the story (Chatterjee and Hotton, 1986). Hallam (1972) has written that the Cretaceous reptiles and Tertiary mammals are basically southern, but Chatterjee and Hotton (1986) pointed out that this is incorrect (e.g.,

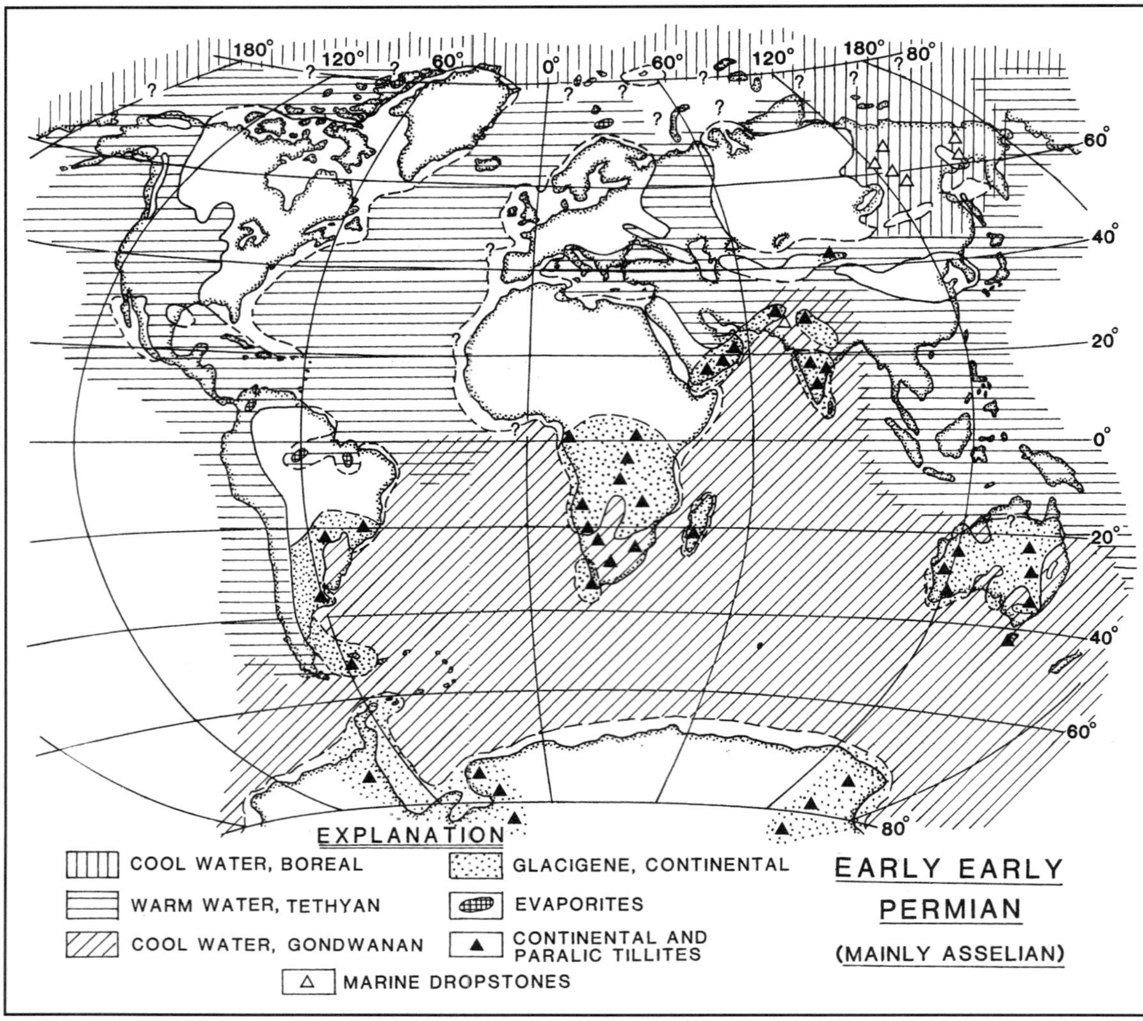

Figure 8. Distribution of temperate and warm-water seas during early Early Permian (mainly Asselian) time.

Colbert, 1938; Dehm and Oettingen-Spielberg, 1958; Pascoe, 1959; Sahni and Misra, 1972; Sahni and Khare, 1973; Gayet et al., 1984; Chatterjee and Hotton, 1986). The Jurassic and Cretaceous tetrapods of India are distinctly northern with their closest relatives in Europe, North America, Africa, eastern and southeastern Asia, and Xizang (Ingavat and Taquet, 1978; Verma et al., 1979; Gayet et al., 1984; Buffetaut and Ingavat, 1986; Chatterjee and Hotton, 1986).

Southeastern Asia

Pre-Carboniferous. Fossiliferous strata of Cambrian through Devonian ages are known from many scattered localities in southeastern Asia, but all are of Tethyan, or northern aspect (e.g., Wongwanich et al., 1983), with the possible exception of some geographically widespread latest Ordovician Hirnantian faunas, and even these are not the pure southern type faunas (Fig. 3). One must keep in mind, too, that during the Hirnantian interval some southern taxa extended into northern Europe and Gaspé, Québec in Canada (Fig. 2), where they overlie northern types in the Baltic region and Gaspé, as well as in northeastern Asia where the same situation occurs. Fang Zongjie (1993), summarizing the available data, showed that the Cambrian and Ordovician biotas are related most closely to northwestern Australian biotas. Silurian-Carboniferous biotas are closest to Rhenish-Bohemian biotas of Western Europe and the Urals.

Carboniferous. On the Malay Peninsula (Malaysia) and Thailand, Mississippian invertebrate marine faunas are rather cosmopolitan, warm-water, shallow-shelf taxa with close affinities to Europe and, to a lesser extent, to Western Australia and North America (Gobbett, 1968, 1973b; Metcalfe et al., 1980; Metcalfe, 1986; Fang Zongjie, 1993). The closest relations with Australia were along the northern Australian coast (Mamet and Belford, 1968; Fang Zongjie, 1993). Pennsylvanian faunas are closely related to those of Russia, Kazakhstan, Europe, and China (Teichert, 1928). The faunas are still Tethyan (Gobbett, 1968; Metcalfe, 1986; Fang Zongjie, 1993).

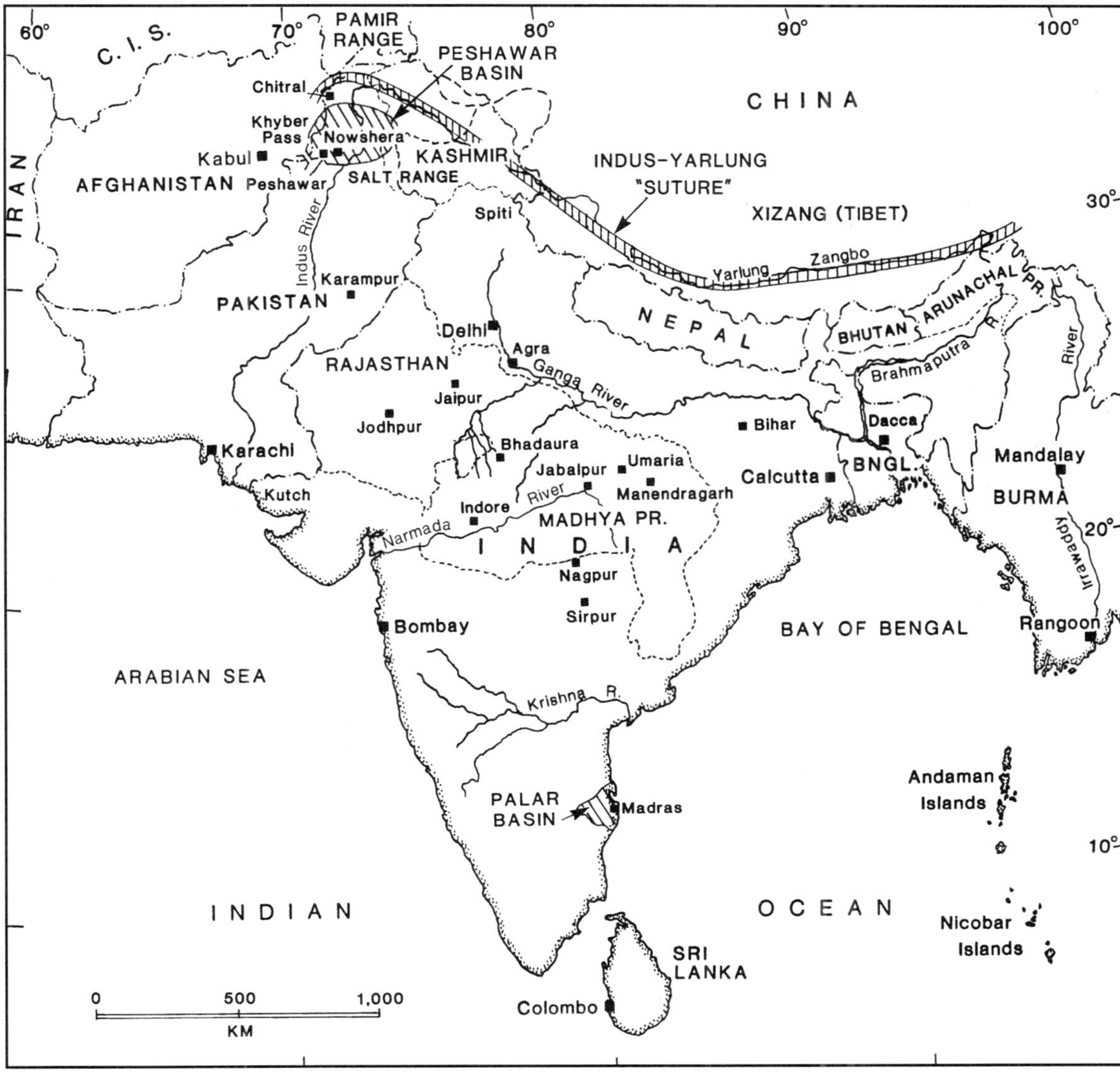

Figure 9. Index map, Greater India, to localities mentioned in the text.

Mississippian to Early Pennsylvanian floras are Euramerian (Euramerican), very similar to equivalent floras collected in Europe, North America, Peru, northern Brazil, Ghana, Morocco, Egypt, Syria, eastern Siberia, China, and parts of Australia (Metcalfe, 1986). A Pennsylvanian flora from Jambi Province, Sumatra (Fig. 1) shows a mixture of elements from the Gondwana flora with elements of the northern Chinese flora, especially with taxa from Shaanxi Province (Fig. 1) (Jongmans, 1937).

Permian. No definite cold-water faunas are known and the whole region may have been covered with warm-water in the earliest Permian (Fig. 8) (Winkel et al.,1983; Dickins, 1985b,d; Dickins et al., 1993). The fauna has been said to show cool-water conditions, but glacigene deposits are probably not present and Tethyan elements predominate (Gobbett, 1968; Grant, 1976; Metcalfe, 1981, 1986; Waterhouse, 1982; Winkel et al., 1983; Dickins, 1985b; Singh, 1987; Fang Zongjie, 1993). In late Early Permian (Fig. 10) and early Late Permian times, warm-water conditions became widespread, with typical Tethyan faunas everywhere, extending to New Guinea and northwestern Australia (Teichert, 1928, 1941, 1951, 1958; Gobbett, 1968; Grant, 1976; Dickins, 1978, 1985b,d; J. R. P. Ross, 1979; Metcalfe, 1981, 1986; Waterhouse, 1982; Winkel et al., 1983; Fontaine, 1986; Archbold, 1987, 1991b). In fact, during Permian time, the whole of modern-day Southeast Asia, Indonesia, New Guinea, and northwestern Australia was a part of the intercalary zone which, in places, was 4,000 km (2,500 mi) wide (Figs. 8, 10, 17). Well-developed land connections existed, however, between Thailand and Indochina in the south, and Eurasia on the north, as demonstrated by the vertebrate fauna (Buffetaut, 1989).

The famous Timor fauna (Audley-Charles, 1968) not only has similarities with the taxa of the Salt Range (Naka zawa, 1985), but also with the fauna of Gondwanaland (Archbold et

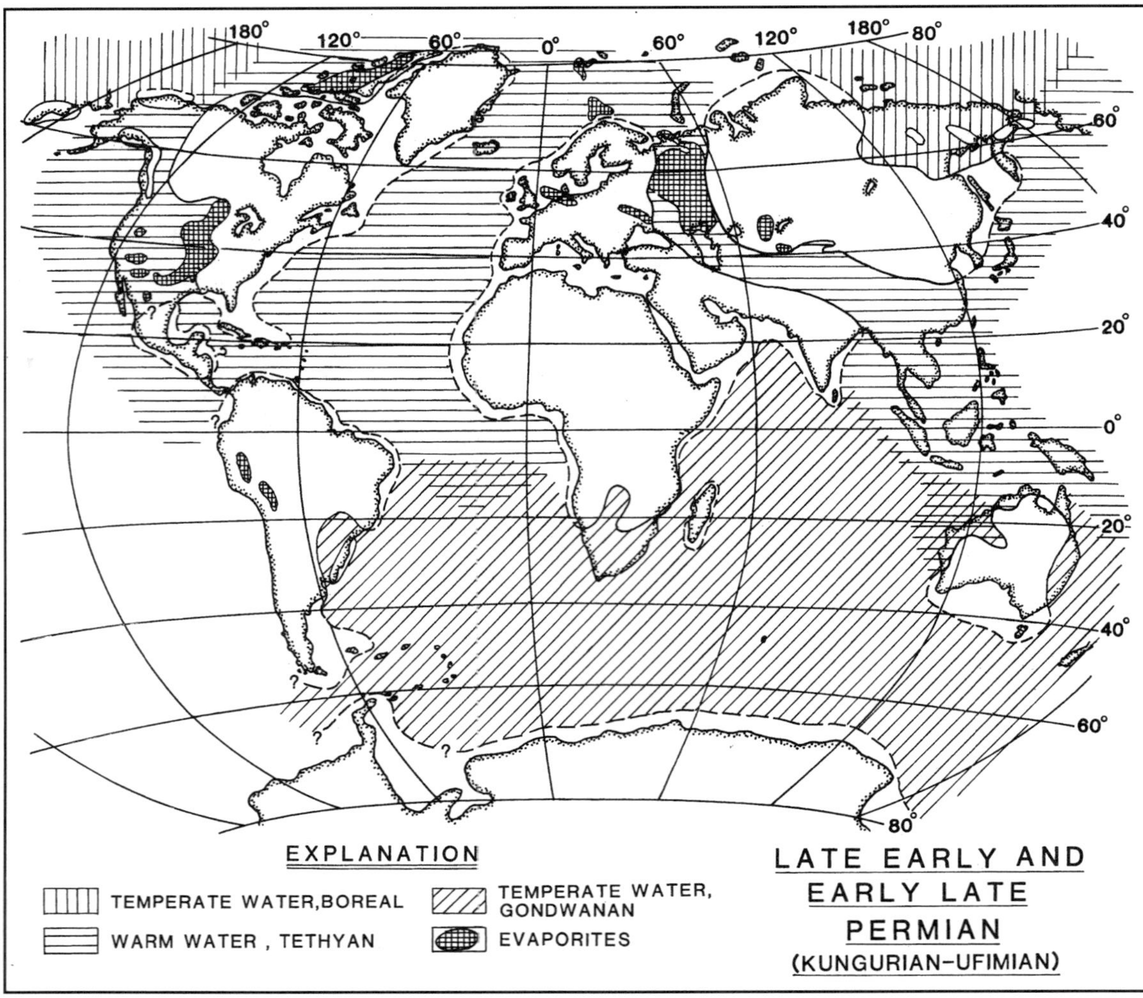

Figure 10. Distribution of temperate and warm-water seas during late Early and early Late Permian (Kungurian-Ufimian) time.

TABLE 1. INDICES OF FAUNAL SIMILARITIES BETWEEN INDIA AND OTHER REGIONS*

Age	Europe	North America	Eastern Asia	Africa	South America	Australia	Antarctica
Early Permian	100	60	0	40	0	0	0
Late Permian	66	33	33	100	33	0	0
Early Triassic	71	0	57	100	0	57	57
Middle Triassic	88	38	63	100	38	25	0
Carnian	75	100	38	63	50	0	0
Norian	100	50	100	50	50	0	0
Early Jurassic	100	25	25	50	50	25	0
Late Cretaceous†	50	100	83	33	33	0	0

*Data from Chatterjee and Hotton, 1986. The values shown are percentages of Indian genera and families present in both India and the region indicated.
†Verma et al. (1979) obtained similar values for the Early Cretaceous.

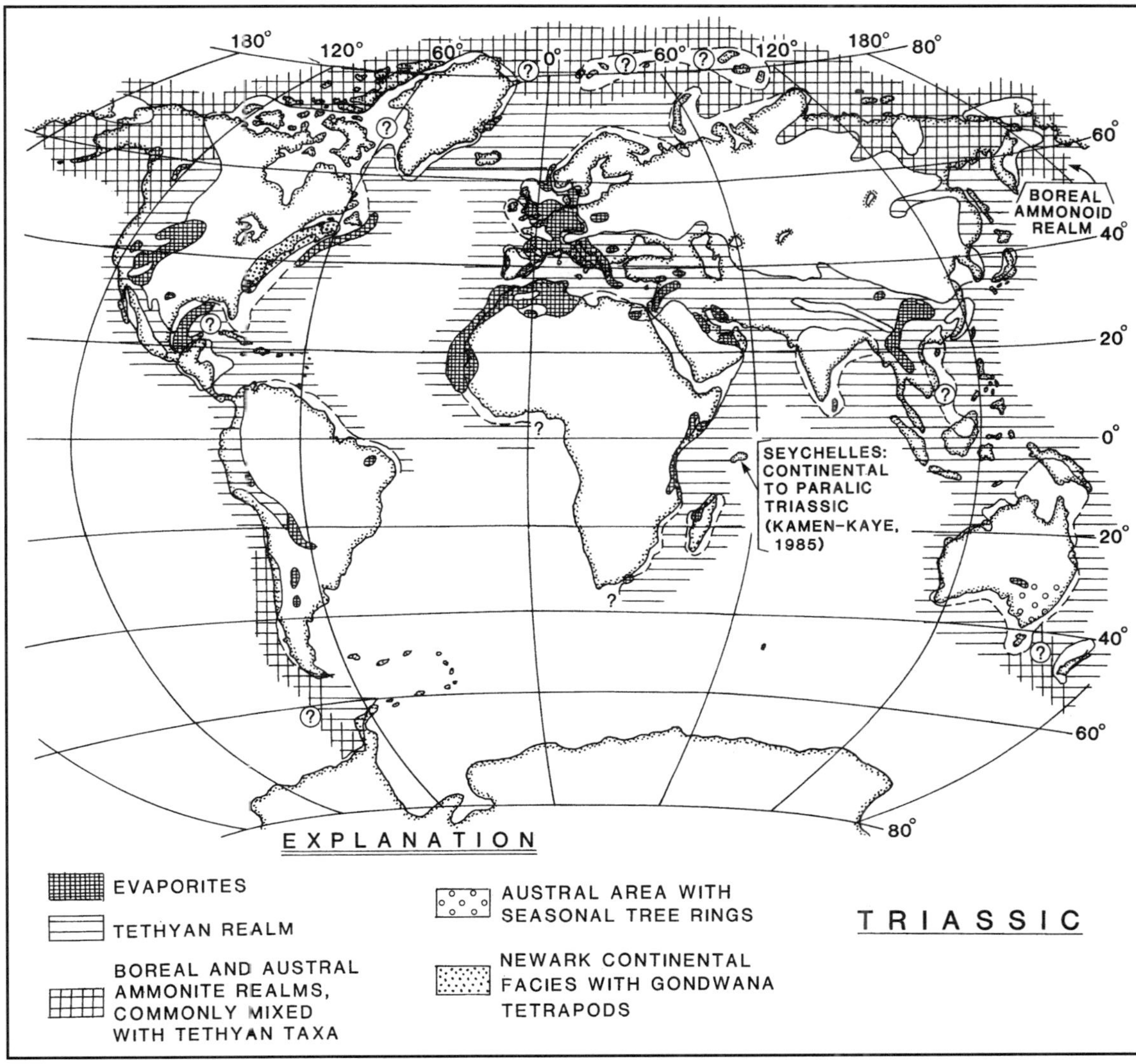

Figure 11. Distribution of temperate and warm-water seas during Triassic time.

al., 1982). Archbold (1981) wrote that the Lower Permian faunas of New Guinea are closely related, even with congeneric and conspecific relationships, to coeval faunas of Timor, Western Australia, Thailand, the Himalaya, and Peru.

The Permian floras of Thailand and Malaysia are mainly Cathaysian (Gobbett, 1968; Metcalfe, 1986). However, Kon'no (1963) described one Permian glossopterid from Thailand. In Indonesia, Plumstead (1973) discussed a Cathaysian flora from Borneo (Fig. 1) that closely resembles a coeval flora from North America. Jongmans (1940) described a Permian Cathaysian flora from southwestern New Guinea, south of the "suture zone" between Australia and southeastern Asia. Close by—only 10 km (6 mi) away—Jongmans (1940) described another Permian flora, this one with *Glossopteris*. Gothan and Weyland (1954) noted that the *Glossopteris* leaf taxa present are typically Australian. Lacey (1975) discussed the possibility of a faunal and floral migration route between Asia and Australia during Permian time. Smiley (1979) concluded that Australia and southeastern Asia had the same geographical relationship during Permian time as they do today, as Teichert had concluded many years earlier on the basis of marine invertebrates (Teichert, 1941, 1951).

Mesozoic (Figs. 11 through 13). Metcalfe (1981) observed that Lower Triassic faunas of the Southeast Asia area are Tethyan, and Chonglakmani (1983) made a similar observation for the whole of the Triassic section of Thailand. Kummel (1969) found that the Scythian ammonoids are Tethyan in the west but become more akin to Pacific faunas farther east. Metcalfe (1990) reported that the entire Triassic section of the lower Malay Peninsula is Tethyan. Archbold (1987) wrote of the Tethyan and cosmopolitan elements of the Middle and Upper Triassic invertebrate faunas of Papua New Guinea. Jurassic and Cretaceous faunas of the region are Tethyan and Pacific, in addition to having cosmopolitan and endemic taxa (e.g., Chonglakmani, 1983). The floras are increasingly modern, with Tethyan forms persisting only in a few large continental blocks. Just

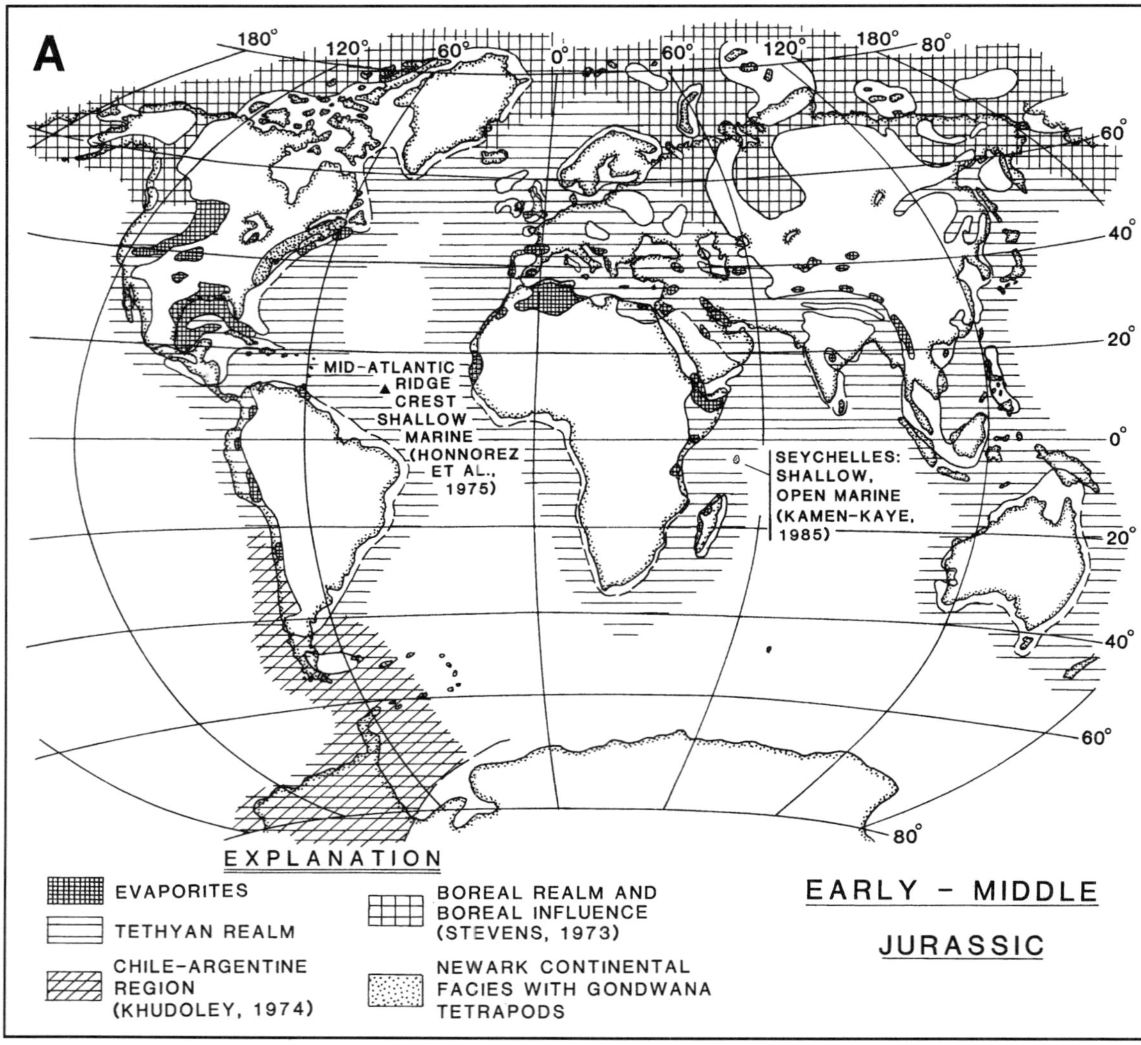

Figure 12. A, Distribution of temperate and warm-water seas during Early and Middle Jurassic time. B (on facing page), Paleogeographical map modified from Drewry et al. (1974) for the Jurassic. Note (by comparison with 12A) that the interpretation of Drewry et al. suffers from combining Early through Late Jurassic data on a single map, since the climatic boundaries for the Late Jurassic are considerably different from those for the combined Early and Middle Jurassic. This results in serious "conflict" between the coal (humid) data and the evaporites (dry) data.

(1952) described a relict *Glossopteris* flora of Triassic age from a locality near the Gulf of Tonkin in Vietnam (Fig. 1). Smiley (1970a,b), describing Late Jurassic–Early Cretaceous floras of Malaysia, noted the close relations among the floras of Asiatic Russia, China, Japan, and India (Indo-european flora of CIS, mainly Russian, authors).

The vertebrate faunas are northern with the sole possible exception of the *Lystrosaurus* fauna and, perhaps, the younger *Cynognathus* zone Triassic fauna. This is true of the faunas of India, Thailand, and Indochina (Chatterjee and Hotton, 1986; Buffetaut, 1989). Reviewing the occurrences of the tetrapods from these two zones, one must consider the possibility that *Lystrosaurus* and its succesors originated in the northern world and migrated southward. As for the Late Triassic, Jurassic, and Cretaceous tetrapod faunas of southeastern Asia, several studies indicate that the taxa studied are mainly northern (Ingavat and Taquet, 1978; Buffetaut and Ingavat, 1986).

Australia

General. The concept of a solid Gondwanaland continent was fairly well entrenched before detailed studies of the fossil floras and faunas of central, western, and northern Australia began. Koken (1907) seems to have been the first to realize that Australia formed an independent geographical unit during much of Phanerozoic time. Schuchert (1935) even called the intercalated marine and nonmarine sections of Western Australia the "Western Australian province," regarding that area as a "last outpost of the Tethyan realm." He interpreted the Western Australian region as a Tethys-Gondwana mixing zone.

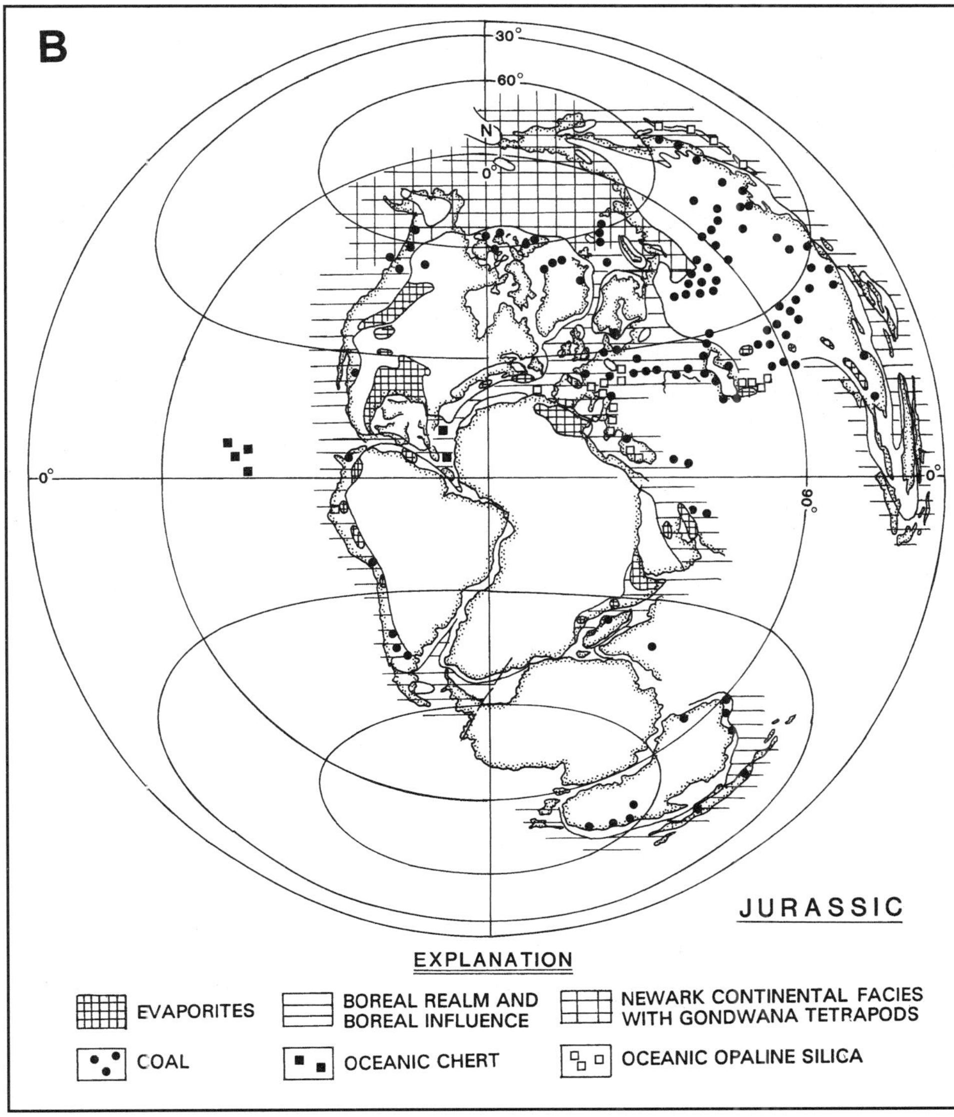

Teichert, who described much of the section in Western Australia for the first time in detail, noted that the fossil record showed clearly Australia's close relationship with Timor and southeastern Asia, and saw no reason to postulate significant changes in that relationship during most of Phanerozoic time (Teichert, 1939, 1940, 1941, 1942, 1943a,b, 1949, 1950, 1951, 1952, 1958, 1972, 1974, 1991). Writing in 1951 (p. 88), Teichert compared Australia's modern marine invertebrate fauna with its contemporaries in Malaysia with their late Paleozoic counterparts. He wrote that:

"The percentage of endemic [living] species of Crinoidea in tropical Western Australia is 66%, of other Echinodermata 50%, of Mollusca 25% and of Brachyura 22%. In spite of close proximity to the eastern Malayan region the character of the northwestern Australian fauna is thus quite distinct. This tropical fauna reaches southward along the coast as far as 29°S. lat., where it gradually merges into the southwestern temperate fauna. It is interesting to note that the limits of its distribution coincide almost exactly with those of the Western Australian Permian faunal province, and it thus seems hardly necessary to postulate significant changes in the relative geographic position of the Timor and Australian regions since Permian time, as already pointed out by the writer in 1941."

Pre-Carboniferous. Strata of Cambrian through Devonian ages are well represented in the Tasman region of eastern Australia, and belong exclusively to warm-water northern-type faunas (Figs. 4 through 7a,b). An early Middle Cambrian trilobite fauna from the northwestern New South Wales part of the Australian platform contains several taxa that are related to taxa in Israel, Siberia, Central Asia, Kazakhstan, and Nevada. The high percentage (35%) of endemic taxa suggests that Australia, or at least its eastern part, was separated from other continents. Evans and Rowell (1990) showed that Australia and at least part of Antarctica formed a single biogeographical unit in Early Cambrian time.

The Australian platform as a whole, in fact, has an extensive

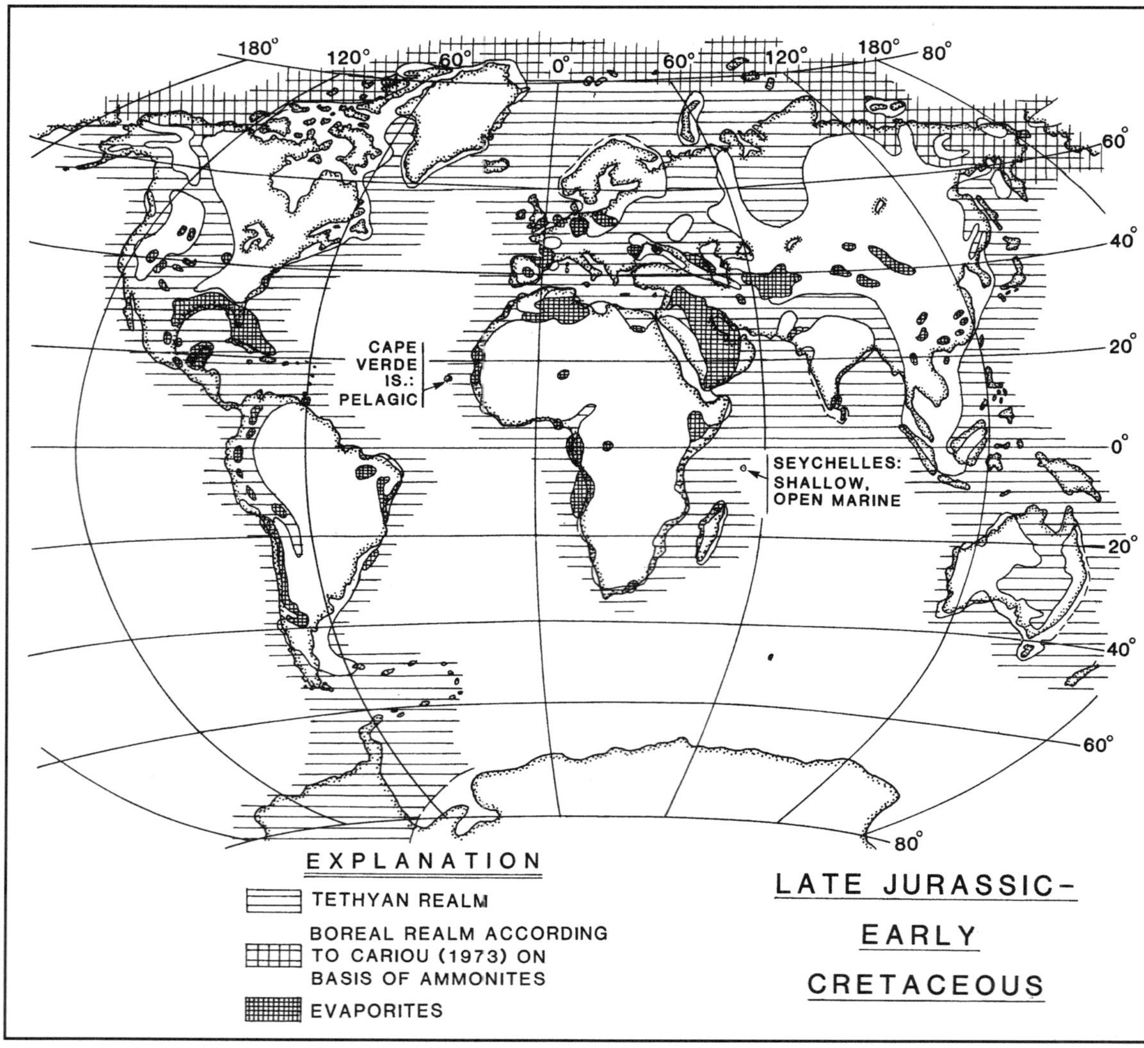

Figure 13. Distribution of temperate and warm-water seas during Late Jurassic and Early Cretaceous time.

early Paleozoic cover bearing fossiliferous northern-type faunas (Teichert, 1943a,b, 1950, 1958, 1972, 1974). Marine Silurian and older Devonian are commonly absent, their places taken in some areas by nonmarine beds. The Early Devonian marine faunas of eastern Australia are endemic at the regional level (Boucot, 1988), whereas the Late Silurian marine faunas of the same region (Rong et al., 1995) are endemic at the provincial level and shared with those of much of southern Asia from Iran through Central Asia and China.

Carboniferous. Teichert (1949) with brachiopods, Hill (1973) and Minato and Kato (1977) with corals, and J. R. P. Ross (1979) with bryozoans found that the northern fauna of the Mississippian extends westward via Tethys to Turkey, and southeastward to Australia. Even though a high degree of endemism characterizes the Australian coral fauna and supports the concept that Mississippian Australia was largely isolated to the same degree as it is today, the large number of cosmopolitan forms present shows that the seas around Australia had direct links to Asia and Europe (Pickett and Wu, 1990). Identical results for foraminifers were found by Mamet and Belford (1968) and Mamet and Playford (1968). Roberts (1985) observed that the Mississippian brachiopod faunas of Western Australia find their closest relatives in Western Europe, Russia, North Africa, and the North American Midcontinent.

Pennsylvanian faunas are more differentiated, with a distinctive Gondwana assemblage appearing (Roberts, 1985). Some ties between eastern Australia and Argentina are evident (Roberts, 1985; Rocha-Campos and Archangelsky, 1985), but a related coeval fauna also is present in the Lake Baykal region of south-central Siberia (Fig. 1), and undoubted links with Southeast Asia also have been demonstrated (Archbold, 1991b).

The trilobites tell a curious story. Engel and Morris (1985) found that, among the several families present in Australia, the Conophillipsiidae have 10 species, with one of them in Japan as well, and 2 in Asiatic Russia. Of the Brachymetopidae, the eastern Australian species also are present in Europe. However, Kobayashi and Hamada (1980) view the Australian Mississippian trilobites as largely endemic.

The Carboniferous floras are more diverse than the Carboniferous faunas. For example, the Tournaisian (Fig. 3) *Lepidodendropsis* flora is very close to that found in Europe and North Africa. In contrast, the Pennsylvanian *Nothorhacopteris* flora is a distinctly southern assemblage, not present in India but possibly related to South America.

Morris (1975) described a Carboniferous section in New South Wales, southeastern Australia, where Viséan (Fig. 3) marine tongues with northern and Tethys-related faunas alternate with continental strata bearing northern plants. About 1,000 m (3,300 ft) above the Viséan marine beds, a *Nothorhacopteris* flora is present, followed 300 m (984 ft) upsection by an Early Permian *Glossopteris* flora.

Permian (Figs. 8, 10). Teichert concluded after years of detailed study of the Paleozoic sections of the Westralian basins of Western Australia that open sea must have lain west of the Australian continent for hundreds of millions of years (Teichert, 1941, 1943a,b, 1949, 1950, 1951, 1952, 1958, 1972, 1974, 1991). He noted that the composition of the faunas—for example, the Permian faunas—suggests a high degree of isolation for Australia. The older Permian strata show cool-water depositional conditions; the younger show warm-water Tethyan conditions (Teichert, 1958; Dickins, 1961, 1973, 1978, 1985d; Archbold, 1987, 1991b; Yeates et al., 1987). These authors, as well as many others (e.g., Minato and Kato, 1977; Dickins and Shah, 1979; J. R. P. Ross, 1979), have commented extensively on the close affinities of the Westralian Permian faunas with those of Timor, Southeast Asia, India, the Himalaya, northern Pakistan, Kashmir, Afghanistan, Iran, Iraq, Turkey, and Europe (Fig. 8). An Early Permian cold-water brachiopod fauna from Tasmania, although largely endemic to the Eastern Australian area, also has affinities with coeval faunas from Western Australia, the Salt Range, and Iran (Clarke, 1990; Archbold, 1991a). Stehli (1973), on the basis of species diversity in brachiopods, placed the southern parts of Australia, New Zealand, Argentina, Chile, and South Africa in a Permian polar region, just as he earlier presented strong evidence for the north and south geographical poles being approximately in their present positions (Stehli, 1968). Ustritskiy (1973) came to an identical conclusion.

The Permian plants also tell an interesting story. Teichert (1958) noted that *Glossopteris*-bearing strata in Western Australia are intercalated in sections bearing the Tethyan marine biota. Miospores of the southern *Dulhuntyspora* assemblage in the Upper Permian, in addition to being present in Australia, New Zealand, and South Africa, have also been found in Antarctica and northeastern India (Venkatachala and Kar, 1990).

Anderson and Cruickshank (1978), studying the relations among the various Gondwana Permian and Triassic tetrapod assemblages, concluded that Australia was relatively isolated by Permian time from the other Gondwana continents, just as Hill (1973) had concluded for Mississippian time. This isolation, noted even earlier by Teichert (1958), probably was complete (four sided). For example, Teichert (1958) reported a Permian marine fauna from Tasmania that is closely related to faunas of the Perth basin (southwestern Australia), and he concluded that open sea lay to the south (Fig. 1). The reality of the isolation is reinforced more strongly by some studies of foraminifers in eastern Australia. For example, Scheibnerova (1981) described 60 foraminiferal species from the Sydney basin. Many of the same species occur also in Western Australia. Most (75%) are cool-water agglutinated forms, and are endemic to Australia, being unknown on any other landmass. Wass (1972a) also has postulated marine seaways between eastern and western Australia.

Mesozoic. Marine Triassic and Jurassic strata are found mainly in the coastal basins and shelves that ring the continent on the west, north, and south, basins that open toward today's coastlines and shelves (Figs. 11 through 13). (This fact alone suggests that Australia was almost surrounded by the sea.) Lower Cretaceous rocks, in contrast, are more widespread, having been deposited far into the interior during the Aptian-Albian transgression of Australia (Fig. 13) (Brown et al., 1968).

Marine Triassic faunas and microfloras show clear and close ties with northern taxa and, in fact, are essentially Tethyan outside of the endemic forms (Dickins and McTavish, 1963; Brown et al., 1968; Balme, 1969; Skwarko and Kummel, 1974; Teichert, 1974; McTavish, 1975; Ash, 1979; Dickins, 1985b,c; Archbold, 1987; Helby et al., 1987; Yeates et al., 1987). Ash (1979) noted that numerous Boreal taxa are intermixed with Gondwana taxa in the same beds. Balme (1969) reported an Upper Triassic microflora in the Carnarvon basin of Western Australia that contains several Madagascan taxa—several at the generic, and a few at the specific level.

Jurassic and Lower Cretaceous faunas and floras show the same phenomena (Yeates et al., 1987). Balme (1969) demonstrated the extremely close relations between the Australian microfloras and those of the Salt Range, Kashmir, and northern India. Some Rhaetian taxa (Fig. 3) are closely related to forms described recently from western China (Zhang Lujin, 1983b).

We have discussed the Triassic vertebrate fauna, which consists mainly of amphibians and relatively few reptiles (in contrast to the largely reptilian compositions in other Gondwana continents), from which Thulborn (1986) concluded that Early Triassic Australia was isolated. We have mentioned the fact that Teichert as long ago as 1958 had recognized Australia's isolation during Mesozoic time from other Gondwana landmasses, despite the fact that, in 1958, the known fauna was small (see Teichert, 1939, 1940, 1941, 1942). Teichert's (1958) conclusions have been corroborated extensively (Anderson and Cruickshank, 1978; Camp and Banks, 1978; Battail, 1981; Warren, 1982; Thulborn, 1986; Molnar, 1992). Cosgriff (1965, 1969, 1972, 1974) and Cosgriff and DeFauw (1987) have made detailed studies of Australia's tetrapod fauna and have firmly established the close affinities of Australia's amphibian population with numerous genera from Western Europe, the Commonwealth of Independent States, India, the southwestern United States, and other northern (not southern) regions. The description of Warren et al. (1991) of Australian Early Cretaceous labyrinthodonts further emphasizes the unique character of the

Australian Mesozoic vertebrates. Finally, the unique little precious opal lower jaw of an Aptian (Fig. 3), ornithorhynchid-like monotreme from Lightning Ridge, New South Wales (Fig. 1), associated with varied dinosaur, crocodile, lungfish, turtle, plesiosaur, and other remains, despite its rarity, provides further support for Australian Mesozoic terrestrial isolation that has persisted through the Cenozoic (Archer et al., 1985).

New Zealand

Pre-Carboniferous. The pre-Carboniferous rocks of New Zealand occur in scattered bits and pieces within South Island. Included are Cambrian, Ordovician, Silurian, and Early Devonian occurrences. None of these is of southern aspect (i.e., Malvinokaffric Realm; Figs. 4 through 7a,b). However, part of the Devonian, the Reefton Beds, does include two Malvinokaffric Realm brachiopod taxa that suggest proximity to the realm (Boucot, 1988). Several taxa imply a close relationship with eastern Australia (Tasman Region of the Old World Realm), but the number of endemic taxa show that New Zealand was sufficiently isolated for provincialism to develop (Boucot, 1988).

Carboniferous. Grant-Mackie (written communication, 1988) pointed out that "The Carboniferous of New Zealand is certainly known at only one very restricted locality (Jenkins and Jenkins, 1971) - based on conodonts in a tectonised marble in the Torlesse zone [Fig. 1]. Lavas and pyroclastics underlying Permian strata in the Nelson-Southland areas have been suggested as possibly Carboniferous, but there is only their deduced stratigraphic position as evidence and some have subsequently produced Permian fossils." Hence to date the rocks do not provide us with much helpful information pertaining to Gondwana and Tethys. Permian rocks, on the other hand, do contain some useful information, as does the overlying Triassic sequence.

Permian. New Zealand records a thick sequence of terrigenous clastics interbedded with volcanics and carbonates (Suggate, 1978). The Early and early Late Permian faunas of New Zealand are close to the generally cool-water marine faunas of eastern Australia, whereas the youngest fauna, which may be of Dzhulfian age (Fig. 3), appears younger than any marine fauna of eastern Australia and shows relations with the warm-water Tethyan faunas of Western Australia (Dickins, 1984). Waterhouse's (1964) cold- and warm-water stages appear to have little or no factual basis and only the apparently youngest fauna (Dickins, 1984, 1985b) has Tethyan affinities as distinct from Australian (Gondwana) affinities.

The single Permian locality in North Island (Suggate, 1978) is a middle Permian, verbeekinid-bearing limestone with strong Tethyan affinities. Two areas in South Island, likewise yielding verbeekinid-bearing limestones, also have strong Tethyan affinities (Gobbett, 1967; Minato and Kato, 1977). Grant-Mackie (written communication, 1988) noted that these "limestones" are not a single unit but merely rather small restricted masses, in one case a float cobble; in others the limestones are small lenses associated with volcanics. These occurrences were interpreted by Spörli and Gregory (1981) as exotic, or displaced terranes, brought in by a "conveyor-belt" convective cell. Archbold (1983, 1987) reviewed the pros and cons of allochthony versus autochthony for these carbonates, finding arguments on both sides. Dickins (1985b) noted close similarities with Australian units and concluded that, at the time these carbonates were deposited, warm-water Tethyan conditions could have extended as far south as New Zealand. We return to the subject of the widths of past climatic zones near the end of this volume.

Gondwana plants are present in the Permian of New Zealand (Plumstead, 1973), and include rare *Glossopteris* (Mildenhall, 1970).

In New Caledonia, Late Permian faunas have a distinct circum-Pacific and Tethyan affinity (Waterhouse, 1969).

Mesozoic. Stevens (1980) has summarized the biogeographical relations of the New Zealand marine fauna. He placed the Triassic and Lower Jurassic in a Maorian Province, as others did earlier, which includes the New Zealand and New Caledonian endemic fauna (Fig. 1). Thus, the apparently endemic aspect of the Maorian Province is still a problem in terms of whether it is merely a South Temperate–marine biogeographical unit that might have included parts of South America, South Africa, and Antarctica, or whether it is truly an endemic New Zealand–New Caledonian unit. Grant-Mackie (written communication, 1988) also pointed out that the spore/pollen floras of eastern Australia are similar to those of New Zealand–New Caledonia.

Stevens (1980) emphasized that the post–Early Jurassic of New Zealand has essentially a Tethyan-type fauna, accompanied by Indo-Pacific benthic organisms. The Lower Cretaceous, in contrast, has a Palaeoaustral (South Temperate) aspect rather than a Tethyan one. Stevens (1980) also emphasizes the importance, during the Mesozoic, of correctly identifying the Torlesse and Murihiku geological units (Fig. 1), because the faunas included in them during the Mesozoic have distinctive biogeographical aspects. The Torlesse is more allied to the Marie Byrd Land region of Antarctica and the Murihiku is more closely allied to eastern Australia. Suggate (1978) provided good background geological and paleontological-stratigraphical data for these problems. Quilty (1977, 1983) has provided additional support for Stevens' conclusions regarding certain later Jurassic bivalves, as has Crame (1981, 1982, 1986, 1987).

Regarding Triassic strata, Archbold (1987) stated that New Zealand and New Caledonia have cool-water brachiopods. Kummel (1969) placed the Scythian (Early Triassic) faunas of New Zealand in a warm-water Western Pacific province, together with coeval taxa on Timor. Our present information suggests that New Zealand's higher level of modern and Cenozoic provincialism is mainly a post-Mesozoic development, particularly with regard to the marine fauna. Not enough is known about its Mesozoic marine and nonmarine vertebrates to provide much meaningful information.

Helby et al. (1987), who found the dinoflagellate *Sverdrupiella* in New Zealand, commented on its Tethyan origins, and mentioned the presence of the same genus in Arctic Canada,

Alaska, Indonesia, and northwestern Australia. Late Triassic to Late Jurassic Radiolaria have been discovered in green and red cherts on North Island, New Zealand. Late Triassic and Early Jurassic Radiolaria have "low-latitude" or Tethyan affinities (Spörli et al., 1989). Middle and Late Jurassic Radiolaria appear to have higher latitude origins, although more study of these assemblages is needed to confirm the results. Retallack (1987) described Triassic floras, noting that these, unlike coeval faunas, are Gondwanan in aspect.

Younger strata show less southern influence, and the vertebrate faunas, with one exception in all the Mesozoic, are marine (Speden, 1973; Keyes, 1977; Molnar, 1981). If this fact holds true as more discoveries are made in the future, it suggests strongly that New Zealand was largely isolated during the Mesozoic.

FAUNAL AND FLORAL REALMS: FROM THE SUBCONTINENT TO THE AMERICAS

Southwestern Asia-Arabia

Pre-Carboniferous. From Papua New Guinea to Africa there are varied occurrences ranging in age from Cambrian through Devonian (Figs. 4 through 7a,b). None of these is of southern aspect except for some widely scattered occurrences of latest Ordovician, Hirnantian-type, Malvinokaffric Realm, shelly faunas correlated with the heightened global climatic gradient of that time (Boucot et al., 1988; Rong and Harper, 1988). As we now know, latest Ordovician time included widespread continental glaciation (Fig. 5) (e.g., Beuf et al., 1971; Vaslet, 1990; and many other references). Hirnantian-type shelly faunas correlate with the heightened global climatic gradient of that time.

The intercalary relations between northern and southern biotas on the Arabian Peninsula are not confined to the Permian section, but are also observed in the older Paleozoic rocks. The Ordovician of Arabia yields a typical Malvinokaffric Realm shelly fauna, including the trilobite *Neseuretus* (Massa et al., 1977; El-Khayal and Romano, 1988). Above that are the latest Ordovician tillites, followed upward by graptolitic Early Silurian beds. This Silurian may be Malvinokaffric, but we cannot be certain because our knowledge of graptolites is insufficient to distinguish Malvinokaffric Realm forms from extra–Malvinokaffric Realm forms. However, its higher land plant spore tetrads of Llanvirnian through Llandoverian age (Fig. 3) are of Malvinokaffric Realm type (J. Gray, 1992, written communication). However, lying disconformably above the Early Silurian is an extra-Malvinokaffric "northern" biota—a warm-water, late Early Devonian marine sequence with a Rhenish-Bohemian Region fauna (Boucot et al., 1988)

Carboniferous. The Mississippian of southwestern Asia contains abundant warm-water (Tethyan) taxa, whose presence indicates the existence of a tropical climate similar to that which we noted from Papua New Guinea and Western Australia to the Salt Range of northern Pakistan (Fig. 1). Detailed faunal lists for the Afghanistan area show that this area and adjacent Iran belong to Eurasia (Weippert et al., 1970).

Permian (Figs. 8, 10). Here, as in so much of the region, the basal unit (Early Permian) is a terrigenous clastic unit. Its fauna includes Gondwanan and Tethyan elements, and can be identified and correlated from Papua New Guinea–Western Australia to Saudi Arabia–Turkey (Fig. 8). This "marginal Gondwana" biota and lithofacies (Hudson and Chatton, 1959; Hudson, 1960) everywhere underlies typical Tethyan facies that range in age from late Early Permian through Aptian or Albian. The sequence in Afghanistan was described by Weippert et al. (1970), de Lapparent et al. (1971), Schreiber et al. (1972), Termier et al. (1974), and Nakazawa (1985). The sequence in Turkey was described by Wagner (1962) and confirmed by Lacey (1975), among others.

The Arabian Peninsula sequence, particularly the critical section in Oman, was described by Hudson and Chatton (1959), Hudson and Sudbury (1959), and Hudson (1960); the megafaunas were studied by Dickins and Shah (1979). The Tethyan coral faunas were studied by Minato and Kato (1977) and Fontaine (1986). In the Hawasina nappes of the Oman Mountains, De Wever et al. (1988) found Tethyan Permian Radiolaria in pelagic sequences within the Hamrat Dura Group. El-Khayal et al. (1980) described a mixed European and Cathaysian flora from the early Late Permian Khuff Formation (Tethyan facies) in central Saudi Arabia. In southern Oman, petroleum has been produced from diamictites in the Early Permian terrigenous-clastic sequence. The presence of tillites in Oman and Saudi Arabia is certain (McClure, 1980; Murris, 1980; Braakman et al., 1982; Kruck and Thiele, 1983). A mixed Cathaysian and Indoeuropean flora is present in Iraq, also without Gondwana taxa (Cyrtoký, 1973). Farther east, however, Gondwana fauna extend well into the Pamir Range of Tajikistan in the former southern USSR (Figs. 1, 9) (Grunt and Dimitriyev, 1973; Nakazawa, 1985).

In Pakistan, Kashmir, northern India, and Afghanistan, Talent and Mawson (1979) made a study of several Devonian and Permian faunas on both sides of the Zagros and Indus-Yarlung suture zones (Fig. 1). They tried, through the use of Simpson and Jaccard coefficients, to show the lack of similarity between coeval faunas of far northern Pakistan (Chitral area) and north-central Pakistan (Nowshera-Khyber-Peshawar area), 200 km (125 mi) away (Fig. 9). They did indeed show significant differences between the two assemblages, but they also showed close similarities. Their study poses several problems:

1. The authors eliminated from their statistics 12 cosmopolitan genera.

2. They did not consider either lithofacies differences or the fact that they were comparing different level-bottom communities.

3. They used data for only one (perhaps two) localities south of their Indus-Yarlung suture zone, so that the sample south of the suture is too limited.

4. They selected the Himalaya Main Boundary Thrust as the suture—which we refer to as "their" suture zone—rather than the now-accepted fault zone farther north along the Indus-

Yarlung "suture." By shifting the position of the suture, they were able to place the Himalaya Tethys localities north of "their" suture.

5. They explained the good correlations between the Peshawar area and the localities in the Tethys Himalaya, Central Asia, and China-Southeast Asia as a consequence of faunas migrating in long circuitous routes *around* Gondwanaland.

6. They made no reference to the monograph of Weippert et al. (1970) on the stratigraphy and faunas, including extensive faunal lists, of central Afghanistan.

7. They wrote nothing on the implications of the Central Afghanistan Channel (Fig. 14; see below).

Mesozoic. The Late Permian and younger section—all of it deposited in Tethys—is too well known for it to merit more than a brief statement. The Tethys was open to the Pacific on the east, into southern Europe on the west, and to East Africa and Madagascar on the south (Figs. 10–13). In the Middle East, including Iran, Iraq, and Arabia, thick evaporites and carbonates are widespread and laterally continuous (especially in the Permian, Jurassic, and Cretaceous), showing that no suture zone requiring thousands of kilometers of tectonic transport (Drewry et al., 1974) ever passed through this region in Paleozoic or Mesozoic time (Kummel, 1969; Kamen-Kaye, 1972; Teichert, 1974; Kashfi, 1976).

Central Afghanistan channel

During Late Cambrian time, a narrow seaway formed across north-central Afghanistan, linking the Paleozoic sea of eastern Iran with that of northern Pakistan, Kashmir, and Spiti in the Indian Himalaya (Fig. 14). The channel, the identification of which is based on many fossil assemblages, (Teichert and Stauffer, 1965; Stauffer, 1968a,b; Weippert et al., 1970; Wolfart, 1970; Pogue et al., 1992) persisted well into Permian time, crosses all of the various plate boundaries that have been proposed to separate Pakistan and India on the east from Afghanistan and Iran on the west. Wolfart (1970) and Weippert et al. (1970) called this seaway the Central Afghanistan Channel (Fig. 14). Weippert et al. (1970), Schreiber et al. (1972), and Wolfart and Kursten (1974) have reviewed many aspects of the channel history. Weippert et al. (1970) published a series of paleogeographical maps for each post-Cambrian Paleozoic system, and gave extensive lists of the marine faunas recovered from the sedimentary rocks within the channel. Teichert and Stauffer (1965), and Stauffer (1968a,b), described Silurian and/or Early to Middle Devonian strata (including reefs) from the channel. Both Gobbett (1967) and Weippert et al. (1970) listed Verbeekinidae (*Neoschwagerina* and associated fauna) from the channel's Permian sector. Pogue et al. (1992) described a nearly complete Paleozoic section from the Peshawar basin of northern Pakistan (Fig. 10). Suneja (1978) traced the channel eastward across the northwestern corner of the Indian craton. The undoubted northern (Tethyan) faunas at all levels and the warm-water character of the litho- and biofacies are emphasized by the presence of thick rock salt and gypsum beds in the Devonian section.

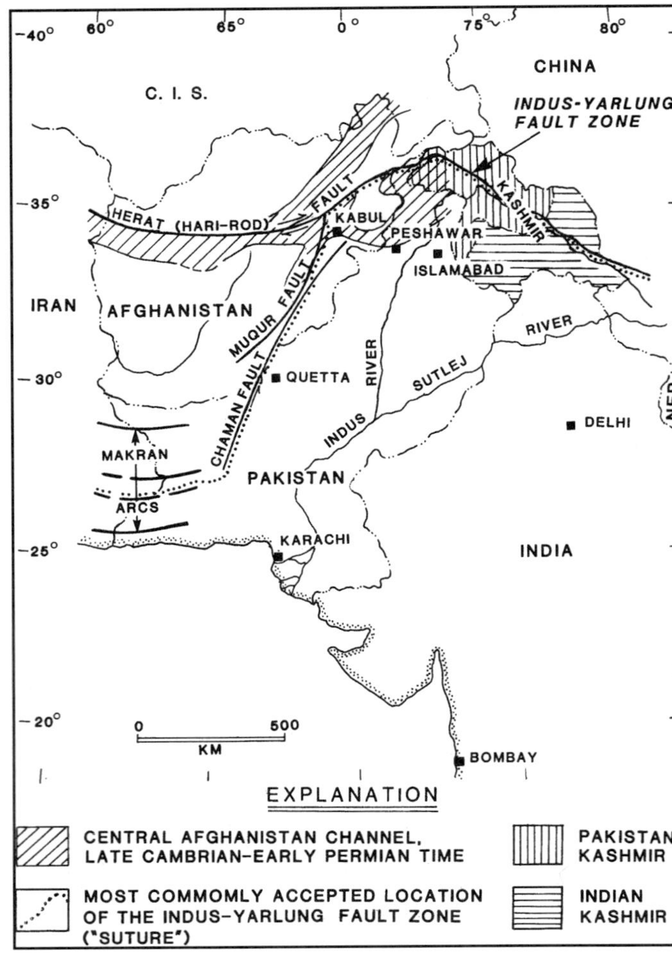

Figure 14. Map showing the location of the Central Afghanistan Channel as mapped by Weippert et al. (1970), Schreiber et al. (1972), and Wolfart and Kursten (1974). This channel crosses all of the post-Paleozoic and pre-Mesozoic sutures postulated in this region essentially intact.

The presence of the channel is a paleogeographically embarassing point that was circumvented by some geologists by placing Pakistan, Afghanistan, Iran, much of Iraq, Saudi Arabia, and Turkey in Gondwanaland (e.g., Weippert et al., 1970; de Lapparent et al., 1971; Schreiber et al., 1972; Chaloner and Lacey, 1973; Termier et al., 1974; Wolfart and Kursten, 1974; Lacey, 1975; Talent and Mawson, 1979; Archbold, 1983). Others, however, have placed the region in Tethys (e.g., Teichert, 1958, 1974; Minato and Kato, 1977; Dickins and Shah, 1979; Waterhouse, 1983; Stöcklin, 1984), as the faunas and floras and lithofacies require. One can either regard this region as a piece of Gondwanaland with Tethyan lithic and faunal character or, conversely, regard it as a Laurasian region having little to do with Gondwana. These differences in interpretation illustrate the futility of using fossils and lithic evidence out of context, and stress the need for an overall synthesis incorporating all available evidence.

Madagascar

Pre-Permian. Fossiliferous pre-Permian rocks are unknown in Madagascar.

Permian. Late Permian marine beds in Madagascar (Fig. 10) contain ammonoids that are identical with, or closely related to, taxa in the Salt Range, Timor, and eastern Greenland; the interbedded continental strata have Gondwanan taxa (Besairie, 1972; Kamen-Kaye, 1978). A seaway, based on both geological and biogeographical data, clearly connected Madagascar with Tethys and Greenland and separated it from Africa (Battail et al., 1987).

Mesozoic. Triassic taxa are even more helpful in reconstructing the paleogeography (Fig. 11). Many are congeneric, even conspecific, with forms in the Salt Range, Kashmir, the Himalaya, Afghanistan, Svalbard, Greenland, Idaho, and British Columbia (Furon, 1968; Besairie, 1972; Kamen-Kaye, 1978; Battail et al., 1987). Balmé (1969), as we have noted, found Madagascan Late Triassic microfloral species in Western Australia. Fish genera of Madagascar also appear in Svalbard, Greenland, Ellesmere Island, British Columbia, China, Nepal, Siberia, and Western Europe (Figs. 1, 2, 9) (Furon, 1968; Besairie, 1972). Associated tetrapods, however, are largely endemic and quite different from those of Africa (Beltan and Tintori, 1981; Battail et al., 1987), perhaps reflecting inadequate sampling. If this observation is real and not a sampling artifact, it shows that Madagascar was separate from Africa.

The Jurassic–Lower Cretaceous succession of ammonite (and other faunal) zones on Madagascar is even more complete than those of Kutch (Figs. 1, 9, 12a,b, 13) (Teichert, 1974). Some of the Jurassic–Lower Cretaceous ammonites are cosmopolitan; many other genera and species also occur in such places as Kutch, Europe (including Great Britain), and elsewhere. Marine Middle and Upper Jurassic ostracodes are conspecific (or at least congeneric) with ostracodes in South Africa and Kutch. Overall, the faunas are diverse and provide no support for most plate tectonic interpretations (Grekoff, 1957, 1963; Furon, 1968; Besairie, 1972; Klinger et al., 1972; Kamen-Kaye, 1978).

The placental mammals of Madagascar have elicited amazement since their initial description long ago (Millot, 1972). They consist almost entirely of an archaic complex of lemurs that is easily conceived of as the descendants of a latest Cretaceous or very early Tertiary set of endemic taxa that have had no reproductive communication with mainland Africa since the Late Cretaceous or early Tertiary. To this must be added the famous Quaternary pygmy hippopotamus that can be explained only as the result of chance transport of an aquatically adapted pregnant female across the Mozambique Channel during the Quaternary. The total absence in Madagascar of the normal African placental fauna demands evolution there in complete reproductive isolation.

We must comment here that, if the Late Cretaceous and early Tertiary vertebrates from Peninsular India—a time when India supposedly was in total isolation as it journeyed from Australia to Asia (Dietz and Holden, 1970; Drewry et al., 1974; Johnson et al., 1976)—showed a pattern of high endemicity like that of Madagascar, we would have thought twice about preparing this volume. However, the evidence from Madagascar for near-total isolation is truly impressive and very difficult to explain as some sort of artifact. India, in contrast, shows no evidence of strong endemicity for Late Cretaceous–early Tertiary time and, in fact, shows overwhelming evidence for its Asian and European geographical connections from at least Cambrian time to the present.

All of the Madagascan data together suggest to us that the area has been in a transitional position at least since the Early Permian. Battail et al. (1987) and Sahni et al. (1987) concluded that the Madagascar vertebrate faunas of the Permo-Triassic and Late Cretaceous–earliest Eocene, respectively, have a Laurasian character (rather than a southern African or highly endemic character) that is not predicted from a "standard" plate tectonic explanation. Battail et al. (1987) and Sahni et al. (1987), on the basis of the near-totally dissimilar tetrapod faunas since Early Permian time, concluded that a water body has separated Madagascar from Africa for at least 256 Ma.

Seychelles Bank

The Seychelles Bank is a continental block in the Indian Ocean northeast of Madagascar (Figs. 11 through 13). Exploration wells drilled on the bank for petroleum penetrated a thick (4,363 m; 14,310 ft) section of continental to paralic Triassic, and open-marine, shallow-water Jurassic, Cretaceous, and Cenozoic (Kamen-Kaye and Meyerhoff, 1980; Kamen-Kaye, 1985). Such results are wholly unpredicted by current tectonic models, which place the vicinity of this continental block some 500 to 1,000 km (310 to 620 mi) from the nearest postulated seaway.

Africa

General. The close bio- and lithofacies relations between southern Europe and northern Africa have long been the basis for the Tethys concept. What is not generally appreciated is the fact that Tethyan taxa dominated more than half of Africa during nearly every time interval since the inception of Tethys (Figs. 4 through 8, 10 through 13).

Pre-Carboniferous. In North Africa, Cambrian faunas are distinctly northern, with carbonate-type faunas dominant in the Early Cambrian (Fig. 4). Recently Early Cambrian marine-carbonate faunas were discovered in West Africa in the southwestern part of the Taoudeni basin and in the Paleozoic Mauritanides fold belt close to the Guinea-Senegal border (Fig. 1) (Culver et al., 1988). In Zambia, Drysdall et al. (1972) discovered Cambrian to Middle Ordovician marine strata along the Luapala River, more than 1,200 km (740 mi) from the present coast.

Ordovician faunas of North Africa are, in contrast, distinctly Malvinokaffric, and reflect the cooling climate of the region, which was glaciated in latest Ordovician time (Fig. 5) (Beuf et al., 1971). In the Ogaden Desert of eastern Ethiopia, Assefa (1988) reported subsurface occurrences of Early Ordovician marine strata with beds of anhydrite. Middle and Late Ordovician rocks are absent. However, the reported presence of Ordovician evaporites, dated by an alleged siphonophore (Assefa, 1988) in a region (North Africa, South Africa) assigned on ample evidence to the cool climate Malvinokaffric Realm, appears unlikely to us.

The Silurian is difficult to characterize for the shelly material is insufficient to be certain, but on the basis of what is known, appears to be Malvinokaffric (Fig. 6). The succeeding warm-climate Devonian is distinctly northern, and North Africa remained free of southern influence (except for its southwesternmost fringe, facing Brazil) again until the Early Permian when cold-climate conditions appeared locally (Boucot et al., 1983; Boucot, 1988; Hiller and Theron, 1988; Figs. 7a,b, 8, 10).

South of Morocco and the Sahara, several marine embayments and basins existed during the early and/or middle Paleozoic. All of them border the present coastal zones, or had open-sea connections with the present coastal and shelf areas (Figs. 4 through 8). The most notable example is the coastal geosynclinal complex of the Rokelide-Mauritanide basins. The Rokelide coastal basin extended from the present vicinity of Tarfaya, Morocco, to Buchanan, Liberia (Fig. 1), a distance of 5,300 km (3,300 mi) (at 5°10'N latitude); it contains marine late Proterozoic to Cambrian strata (Culver et al., 1988; Bertrand-Sarfati et al., 1991). The successor geosynclinal basin, the Mauritanide trough, extends 3,800 km (2,350 mi) from Tarfaya to the Bove basin of Sierra Leone (Fig. 1), and was filled with Ordovician through Middle Devonian strata. The foreland Taoudeni basin opened toward Europe on the north and into the South Atlantic in the south (Figs. 1, 4 through 8; Bertrand-Sarfati et al., 1991). The Cambrian and Devonian faunas are northern, whereas the Ordovician, and possibly the Silurian, are southern, that is, Malvinokaffric Realm (Figs. 4 through 7a,b) (Machens, 1973; Martin, 1982; Wright et al., 1985; Culver et al., 1988, 1991; Bertrand-Sarfati et al., 1991).

Gray (1988) has discussed the palynological data bearing on the Llandoverian, Early Silurian age (Fig. 3) of the Elmina Sandstone occurring on the Ghana shoreline in the Takoradi area. The palynomorphs from the Elmina have much in common with coeval beds present in the Maranhao-Parnaiba basin of eastern Brazil.

Devonian marine and littoral strata are widespread along the African coast from Guinea to offshore Benin (Figs. 1, 7a,b) (Machens, 1973; Martin, 1982; Culver et al., 1991). Martin (1982) and Culver et al. (1991) have described the fossiliferous marine Late Ordovician–Late Devonian sections of the Bove basin (Fig. 1). Machens (1973) identified marine to littoral Devonian in outcrops along at least 200 km (125 mi) of Ghana's coast. Saul et al. (1963), Anderson et al. (1966), and Saul (1967) studied fossils from the Accra embayment in Ghana, and concluded that they are Early or Middle Devonian representing mixed Malvinokaffric Realm and northern elements. The latter occurrences show that marine connections with the Eastern Americas Realm and the Rhenish-Bohemian Region existed. Anan-Yorke (1974) has discussed the Devonian chitinozoans and acritarchs present in several offshore wells from Ghana. Minor subsurface gypsum "shows" are present (Saul et al., 1963; Martin, 1982) but probably are not climatically significant when weighed against their proximity to the cool climate Malvinokaffric Realm fossils.

Of special interest are the data from offshore wells drilled during petroleum exploration. Paleozoic marine strata have been found offshore from Liberia to Benin (Figs. 1, 7a,b) (Martin, 1982). Using surface and subsurface data from on- and offshore, Martin (1982) showed a gradation from nonmarine Devonian sediments onshore to open-marine Devonian sediments offshore near the shelf edge. On the basis of spores and marine fossils (that were not listed or figured), Martin (1982) concluded that the marine Devonian Accraian series of Ghana and its lateral equivalents are very widespread, extending 1,600 km (990 mi) from Liberia to Benin and occupying a minimum area on the present shelf of 70,000 km^2 (27,000 mi^2). Provenance studies led Martin (1982) to conclude that the source of the Devonian terrigenous detritus is the exposed West African (Birim) craton. We cannot comment on Martin's (1982) interpretation since we did not examine his data.

In Namibia (Fig. 1), the Nama Group is a marine sequence that was deposited in shallow coastal waters. It consists mainly of clean, mature sandstones (now somewhat quartzitic as a result of mild metamorphism) and shales, but contains important thicknesses of limestone and dolomite. Trace fossils, including some Ediacaran taxa and the presence of certain lithologies (e.g., Late Proterozoic tillite underlies the Nama Group), led Germs (1972, 1973) to date the Nama as Vendian (Fig. 3), possibly extending into the earliest Cambrian.

In South Africa, the latest Ordovician Soom Shale Member and the Disa Siltstone beds above it (almost on the Ordovician-Silurian boundary) contain Malvinokaffric Realm shelly fauna (Figs. 5, 6). Above is the classic Bokkeveld Group Malvinokaffric fauna (late Early and early Middle Devonian) (Fig. 7a,b), followed in turn by the extra-Malvinokaffric warm-water *Tropidoleptus* fauna in the lower Witteberg Group (Boucot et al., 1983). The brachiopod *Tropidoleptus* is known elsewhere in the Rhenish-Bohemian Region of the Old World Realm, and in the Eastern Americas Realm. The fauna occupies suitable ecological niches around the Atlantic, but is absent in Asia and Australia, indicating that marine dispersal routes were widespread in the Atlantic region during Devonian time. In fact, thus far all North African marine Devonian, except in Ghana (and possibly Guinea-Bissau), is northern (Fig. 7a,b).

Carboniferous. Cosmopolitan and Tethyan marine invertebrate taxa predominated in northern Africa during the entire Gondwana interval of the Earth's history, as described in scores of publications (e.g., Furon, 1968; Mamet, 1972; Minato and Kato, 1977; Legrand-Blain, 1985b; Lys, 1985; Semenoff-Tian-Chansky, 1985). The Carboniferous faunas suggest that Eurafrica was far removed biogeographically from the Americas during Carboniferous time (e.g., Mamet, 1972).

South of the Sahara, Carboniferous marine occurrences are little known. Kamen-Kaye and Meyerhoff (1979) and Meyerhoff (1984) reported the occurrence of Mississippian paralic to marine strata in offshore Ghana. Racheboeuf et al. (1989) described a fully marine Mississippian faunule from the Ghana coast at Sekondi. Martin (1982) noted that the offshore Carboniferous section is cut out by an unconformity overlain by Permian(?) nonmarine strata east of the longitude of Accra. West of Accra, the Carboniferous section (which reaches more than 400 m (1,310 ft) in thickness) consists of alternate marine and nonmarine strata, including sandy limestone and dolomite. Unfortunately, faunal studies have not been made of the marine taxa (or, if studies have been made, the results are unknown to us). The fauna is younger than Devonian and is pre-Permian (J. A. Momper, oral communication, 1984). The area known to be covered by Carboniferous strata in the offshore exceeds 20,000 km^2 (7,600 mi^2).

In northern Namibia, a well drilled in the Etosha pan (part of the Okawanga basin centered in Angola) during 1969–1970 recovered, just below Mississippian continental beds, a marine section yielding (J. A. Momper, oral communication, Oct. 15, 1984) conodonts of possible Cambrian age (the conodont, actually a single fragment, was concluded by Sweet [written communication to A. A. Meyerhoff, 1973] to be probably of Devonian or younger age, whereas Momper [written communication to A. A. Meyerhoff, 1984] concluded that regional relations indicated a Cambrian age.).

Farther south in South Africa, a marine incursion in the Witteberg Group was described by Theron (1962) and Gardiner (1969). Gardiner (1962, 1969) mentioned the presence of marine pelecypods, fish, and a few other taxa in the upper part of the Witteberg. A basal Witteberg marine incursion is late Middle Devonian to early Late Devonian (Boucot et al., 1983).

Visean floras are cosmopolitan, and are essentially the same from Niger to Morocco (Fig. 1) (Lejal-Nicol, 1985). The same is true of Visean microfloras (Coquel, 1985). Tournaisian floras in Egypt show some mixing with Argentine, Chilean, Himalayan, Mongolian, and Siberian genera; the corresponding microflora is very provincial. In southwestern Egypt, several taxa closely related to southern forms are present, similar to coeval floras in Morocco, Ghana, and India (Klitzsch and Lejal-Nicol, 1984; Legrand-Blain, 1985b). These facts assume local, even regional, paleogeographical importance during Early Permian time.

Permian. The two known Permian sequences along the western African coast are continental (Jardine et al., 1969; Martin, 1982). The first sequence penetrated in wells drilled offshore from Ghana, consists of fluviatile and lacustrine strata that unconformably overlie Carboniferous marine and terrestrial strata (Martin, 1982). These nonmarine strata are intruded by dolerite dikes and sills that yield radiometric dates of 192 to 176 Ma (Early to Middle Jurassic; Schlee et al., 1974). Although the continental strata may be Triassic or even earliest Jurassic, regional geological relations suggest strongly that the offshore Ghana section is Permian, probably Late Permian.

In Gabon (Fig. 1), deep wells have penetrated continental strata beneath Mesozoic rocks. The oldest of the pre-Mesozoic units is the Agoula Series which consists of sandstone, siltstone, shale, and anhydrite of marine affinities (Jardine et al., 1969; J. C. Hazzard, written communication, 1972). The beds contain *Pemphicyclus gabonensis* (an estherian) and a Dwyka microflora (Jardine et al., 1969; Franks and Nairn, 1973).

In southern Africa, the Dwyka Group of Namibia and South Africa south of about latitude 28°S is dominantly glaciomarine, whereas to the north it is dominantly terrestrial (Visser, 1990), and contains at least two fossiliferous marine incursions, one of which contains a Permian goniatite (Martin et al., 1971). The faunas from these marine incursions are quite diverse and have received increasing attention from biogeographers (e.g., Martin, 1953, 1973; Dickins, 1961, 1985b; Gardiner, 1962, 1969; Hart, 1964; Lane and Frakes, 1970; Martin et al., 1971; Wass, 1972b; McLachlan and Anderson, 1973, 1975; Rust, 1973; Teichert, 1974; Kamen-Kaye, 1978; Loock and Visser, 1985; Oelofsen, 1987). The marine collecting localities are sufficiently widespread that we can state that at least 180,000 km^2 (69,000 mi^2) of the Great Karoo basin, all 9,000 km^2 (3,500 mi^2) of the Warmbad basin, and 42,000 km^2 (16,000 mi^2) in the western part of the 250,000 km^2 (96,000 mi^2) Kalahari basin—231,000 km^2 (88,800 mi^2) in all—were covered at times by shallow-marine (or slightly deeper) Early Permian waters (Fig. 8). The faunas include mesosaurids (aquatic reptiles), goniatites (a poorly preserved specimen, originally described as a nautiloid, but later recognized as an ammonoid, probably a ceratite, and assigned by Teichert and Rilett, 1974, to *Paraceltites*), gastropods, bivalves, brachiopods, foraminifera, marine decapods, scolecodonts, sponges, echinoderms, bryozoans, and radiolarians. Many are endemic, including the more brackish items, but a surprising number—including the bryozoans—have been described as congeneric, in some cases conspecific, with taxa in North America, the former USSR, Kashmir, Spiti, the Salt Range, Timor, Australia, and New Zealand (Dickins, 1961; Lane and Frakes, 1970; Wass, 1972b; McLachlan and Anderson, 1973, 1975). If the specific and generic determinations can be taken at face value, the seas in this part of Gondwana had reproductive communication with similar taxa elsewhere and access to other oceans. The same conclusion was reached by Oelofsen (1987) on several other grounds: (1) the marine embayment in the Great Karoo basin opens toward the west, nearly opposite the Parana basin of southeastern South America, also with mesosaurids; (2) except for two of the three species of mesosaurids, there is an overall dissimilarity between the faunas of the Karoo and Parana basins; (3) the brachiopods of the Great Karoo basin include several taxa with a northern affinity; and (4) the presence of marine evaporite-bearing strata demonstrates access to the world ocean.

In somewhat younger Permian strata of the Parana basin, Runnegar and Newell (1971) found a rich marine fauna (probably brackish, rather than normal marine salinity), which they concluded was endemic to the Parana basin. Runnegar and

Newell (1971) emphasized that this fauna reminded them in its relatively low diversity of Caspian faunas, which yielded a large number of endemic genera among the bivalves. Cooper and Kensley (1984), however, found a very similar fauna in the Ecca Group of southern Africa; at least one taxon is conspecific and others are apparently congeneric. Cooper and Kensley mentioned some related forms from Australia. So few collections of invertebrates have been made in this part of the world that we believe that the Cooper and Kensley (1984) discovery is the forerunner of many more such finds. M. R. Cooper (written communication, 1992) mentioned a "rather poorly preserved bivalve from the Dwyka/Ecca of Namibia which may be part of this fauna." The fact should be noted that these fossils occur in embayments along the *present* coastlines, embayments that had two or more marine incursions during late Paleozoic time from the direction of the *present* oceans (Fig. 8).

Farther northeast in Africa a marine band, or layer, containing Permian acritarchs has been found in a well bore in western Zambia, 1,250 km (775 mi) from today's ocean; the marine band is near the base of the Dwyka (Drysdall et al., 1972). M. R. Cooper (written communication, 1992) suggested that any marine bands in the Zambian Dwyka would most likely have been linked with northwestern Zimbabwe and the Kalahari Basin of Namibia (Fig. 1) (Oesterlen, 1990), he pointed out the difficulty with the relatively unfossiliferous beds involved in distinguishing between lacustrine and marine beds in the absence of useful fossils. Still farther north, opposite Madagascar, Cox (1936) found Permian marine pelecypods in a site 240 km (150 mi) from the present ocean (see also Spence, 1957; Furon, 1968; Kamen-Kaye, 1978), and farther west in southwestern Tanzania, 570 km (350 mi) inland, Wopfner (1991) found marine to paralic Late Permian tongues in the Mhukuru Formation of the Karoo (Permotriassic) Ruhuhu basin near Lake Nyasa (Lake Malawi) (Fig. 1). A marine embayment along the present East African coastal area clearly existed by Late Permian, possibly earlier, time. In support of this last statement, we note the presence in coastal Tanzania of very thick marine evaporites whose oldest layers may extend through the Late Permian (Fig. 11) (Furon, 1968; Kent et al., 1971).

Floras associated with the Early Permian Dwyka Group glaciomarine diamictites and terrestrial tillites are found in two areas of Africa. The larger area is in the south, and includes all of southern Africa south of a line running approximately east-northeast from northern Gabon to northeastern Zaire, just north of the present equator, then south-southeastward to Mozambique (Fig. 1). The northern area, as far as is known, is centered in the area where Egypt, Libya, Chad, and Sudan meet (Fig. 8) (Klitzsch and Lejal-Nicol, 1984; Legrand-Blain, 1985a). Future studies may show that the two areas are connected, but the intervening zone is so far characterized by European-related taxa (Whiteman, 1971; Lacey, 1975). The Dwyka areas are rich in Gondwana flora, and are associated with diamictites (Haughton, 1969; Beuf et al., 1971; Dow et al., 1971).

Mixed floras are known in several places, mostly on the eastern side of Africa. These include northeastern Tanzania (Haughton, 1963; Furon, 1968), northern Zaire (Hoeg and Bose, 1960; Lacey, 1975), Zimbabwe (Walton, 1929; Lacey and Huard-Moine, 1966; Lacey, 1975), Mozambique (Teixeira, 1946, 1947, 1952; Lacey, 1975; Vozenin-Serra, 1984), and Zambia (Lacey and Smith, 1972; Lacey, 1975; Vozenin-Serra, 1984). Maithy (1976) was doubtful about the mixing in the Zimbabwe locality, but Lacey (1975) seemed to think that it is real.

Of considerable importance is the published study by Anderson and Anderson (1985) who, like Chandra and Surange (1979) in India, concluded that none of the leaf form species in Africa is found outside of the African region. Such a statement seems somewhat premature without additional studies of fructifications. However, it is significant that the few fructifications actually studied are so far unknown outside of Africa. If this conclusion holds, the implications are clear; specifically, the continents were as widely separated during Permian time as they are today.

Kovacs-Endrody (1991), however, took exception to the Anderson and Anderson (1985) study; in fact, she took issue with most studies of the South African glossopterid flora made between Seward's (1897) time and that of Anderson and Anderson (1985) nearly 90 yr later. Stating that the observed differences in the glossopterid leaves have genetic significance (i.e., leaf differences, in her opinion, do differentiate biological genera and species), Kovacs-Endrody (1991) selected 25 species of glossopterid leaves from the Transvaal Province in South Africa and attempted to show that they do indeed comprise parts of coeval Brazilian, Indian, and Australian Lower Permian floras. Although she made a strong case for several taxa (e.g., *Glossopteris taeniopteroides* from both Australia and South Africa), her conclusions concerning some other form species are suspect (e.g., *G. browniana* and *G. indica*). Regardless, Kovacs-Endrody's (1991) results do cast some small doubt on the Anderson and Anderson (1985) study and, by implication, that of Chandra and Surange (1979) for India.

On the other hand, Kovacs-Endrody's (1991) results are highly suspect for reasons other than taxonomical errors and the alleged biological significance (or insignificance) of differences in glossopterid leaf characteristics. She either ignored, or attached no significance to, several dangers inherent in correlations with plant form genera:

1. Climate affects leaf morphology profoundly (Meyerhoff, 1952). As a consequence of climate, plant leaves (and other plant parts as well) adapt themselves to forms best suited for their survival in seasonally hostile environments. As Meyerhoff (1952) showed, without adequate preservation of the finest detail in leaves, plants of nearly identical leaf shape and major venation are (and have been) easily mistaken for one another. Meyerhoff (1952) provided several illustrations of this among members of the birch (Betulaceae), maple (Aceraceae), beech (Fagaceae), witch hazel (Hamamelidaceae), walnut (Juglandaceae), willow (Salicaceae), linden (Tiliaceae), and elm (Ulmaceae) dicotyledon families.

2. Some plant genera are of polymorphic leaf habit. For example, the grown leaves of a juvenile American poplar, the common cottonwood (*Populus fremonti*), are indistinguishable from those of several species of willow (*Salix*) of the same family. Polymorphic leaf habits are fairly common in the plant kingdom (Jepson, 1925).

3. Kovacs-Endrody (1991) did not consider the very common problem of climatic zonation by elevation and by latitude (Axelrod, 1944). Plants fossilized at one latitude, if correlated with identical taxa that were fossilized at another latitude 1,500 km (900 mi) away (e.g., Pliocene versus Holocene localities in Canada and Greenland) (Funder et al., 1985), can be miscorrelated by periods in time up to 10 or 15 m.y. Similarly, fossils of a particular flora found at different places at the same latitude need not be correlative. The oldest species of the fossil flora may have been swept in from adjacent highlands and buried. Later, as a consequence of a cooling Earth climate, the same flora may grow at the same latitude but at sea level (Axelrod, 1944).

4. Still another factor not considered by Kovacs-Endrody (1991) is the problem of disjunct versus relict floras. Disjuncts may or may not be correlative. Relicts can be a nightmare in correlation, because it is not always possible, in terrestrial strata to determine how many millions of years a particular flora may have been a relict (Cain, 1944).

A dramatic example of how misleading leaf form can be is the discovery of glossopterid leaves in the Middle Jurassic of Oaxaca State in southern Mexico (Delevoryas, 1969; Delevoryas and Person, 1975). Delevoryas and Person (1975), while noting the uncanny resemblance of the leaves to four known leaf species of *Glossopteris* (*G. indica*, *G. browniana*, *G. taeniopteroides*, and *G. euryneura*), concluded that the leaf species is wholly unrelated to *Glossopteris* because the associated flora is known only from the Jurassic (e.g., *Zamites*, *Otozamites*, *Pterophyllum*, *Ptilophyllum*, *Taeniopteris*, and several other well-known Jurassic taxa).

Finally, it is apparent that Kovacs-Endrody's (1991) underlying taxonomic approach is excessively typological. She assumed without justification that just about any morphological differences shown by specimens assigned to the same genus have specific value. The evidence against this assumption for just about all carefully studied Phanerozoic organisms is overwhelming, although it was an all too dominant attitude in many quarters during the 19th and the first half of the 20th centuries. Variation within taxa is a biological fact of life that all paleontologists must consider if their specific level work is to be taken seriously.

Regarding tetrapods, Romer (1973, p. 167; see also Cooper, 1980) wrote that the Permian reptiles studied at that time made it ". . . quite certain that there was easy communication between Russia and South Africa in both Middle and Late Permian, despite the presumed intervention between the land areas of a Tethys Sea." For example, the middle Permian therapsid fauna of South Africa (Beaufort, Ecca Groups) is closely related to that in Russia (Barry, 1975). Similarly, the Late Permian genus *Diictodon* is present not only in the Cape Province of South Africa but also in the Xinjiang region of northwestern China (Fig. 1) (Sun, 1973b; Cluver and Hotton, 1979). Even the fish faunas from the Permian of Kenya have species in common with those of correlative Russian strata (Furon, 1968).

Triassic. Triassic marine invertebrates are known from two areas. The first is the East African coastal zone from Madagascar to Tanzania north to Kutch, western India (Figs. 9, 11). Here are shallow, brackish-water sediments and thick evaporites (Kent et al., 1971; Foster et al., 1994). Similar and partly coeval marine Triassic salts in western Africa extend from Tunisia (Busson, 1970) and Egypt (Norian-Rhaetian evaporites of the Fadda Formation: Keeley et al., 1990) to Senegal (Ayme, 1965), and belong to the Tethys.

The Triassic vertebrates show that connections existed among all of the continents at one time or another during the Triassic. Australia was relatively isolated during much of Permian-Mesozoic time, and especially during Early Triassic time as documented by Thulborn (1986). South America also seems to have been partly isolated (Chatterjee, 1986a), but this observation may be an artifact of collecting. Hammer et al. (1990) presented evidence suggesting that Antarctica may have been isolated, especially after *Lystrosaurus* Zone time. Chatterjee (1986b) has shown that, for Late Triassic time, communications among the continents were excellent. Three congeneric forms lived in Morocco and Texas (Fig. 2); at least one, possibly three, lived in South Africa and Texas; and four lived in India and Texas. The first Triassic ornithischian dinosaur ever collected in Laurasia (all the rest are in Gondwana) was found in Late Triassic strata in Texas (Chatterjee, 1984).

Bessaire (1972) noted the great similarities among the Triassic fishes of Madagascar, Svalbard, and Greenland. More recently, Beltan and Tintori (1981) studied the Early Triassic genus *Saurichthys*, and found it in Canada, Greenland, Svalbard, Europe, Nepal, South Africa, Madagascar, and Australia. There can be little doubt that this genus, like so many fossil fish, had an anadromous or catadromous habit. Such geographical diversity makes one think of oceans.

Jurassic-Cretaceous. Jurassic through Early Cretaceous faunas are Tethyan, from Somalia to the shelf south of South Africa (Figs. 12a,b, 13) (du Toit, 1954; Arkell, 1956; Dingle and Klinger, 1971; Klinger et al., 1972; Westermann, 1975; Kamen-Kaye, 1978; Dingle et al., 1983). The flora is also northern, especially beginning in Late Jurassic time (du Toit, 1954), becoming distinctly Wealden by the beginning of Cretaceous time (Fig. 3). Ornithischian dinosaurs show that continental links were almost as well developed during the Jurassic as during the Permian and Triassic Periods (Chatterjee, 1984), with similar taxa present in Lesotho, Portugal, Great Britain, China, Montana, and Texas (Figs. 1, 2). The ornithopod *Dryosaurus* (Upper Jurassic), for example, is present in Tanzania, the western United States, and probably in Europe (Charig, 1971, 1973; Galton, 1977). Thus, as more fossil collections have been made,

Colbert's (1979) statement—that the decrease in the number of vertebrate similarities during Jurassic time reflects the breakup of Gondwanaland—requires revision. However, there is an evident trend toward higher provincialism among the dinosaurs that culminated in the later Cretaceous.

Antarctica

Pre-Permian. Cambrian carbonates are widespread in Antarctica (Fig. 5), and also formed in Africa (Early Cambrian archaeocyathid pebbles in the Dwyka) and Australia (Roberts and Jell, 1990; Wood et al., 1992). Evans and Rowell (1990) concluded that Australia and part of Antarctica comprised a single biogeographical province. Fossiliferous Ordovician and Silurian strata are unknown, except for possibly the very earliest Ordovician (Fig. 6) (Webers, 1970). However, the marine Devonian biota (Horlick Mountains; Fig. 15) is of Malvinokaffric Realm type (Fig. 7a,b). Higher in the section in later Middle Devonian beds the presence of the cosmopolitan fish *Groenlandaspis* associated with calcrete indicates a major climatic change from cool to warm (calcretes suggest warm, semi-arid climate). This warm later Middle Devonian climate was followed by a gradual return to cold conditions that culminated in the widespread Early Permian glaciation. The faunal record of Antarctica demonstrates the long existence there of intercalary cold and warm climates.

A recent study of thelodont scales from Victoria Land (Turner and Young, 1992) indicates a late Middle Devonian (and possibly slightly younger) age for the Aztec Siltstone of the Beacon Supergroup. The thelodonts show close affinities with ones from Iran, Thailand, and Bolivia.

Permian. The continental Permian contains the *Glossopteris* flora, but detailed studies have not been undertaken. The trunks of large trees have been found in Antarctica at very high latitudes but, as demonstrated by Funder et al. (1985), Francis and McMillan (1987), and Burckle and Pokras (1991), the presence of large trees at very high latitudes is not a valid argument for postulating continental movements. Taylor et al. (1992) recently described a Late Permian forest from Mount Achernar in the Transantarctic Mountains, a forest whose stumps have pronounced tree rings and which originally grew near 80 to 85°S. Bose et al. (1989) noted the low generic and specific diversities that would be expected in a cool, polar climate. Although no marine Permian strata have been recognized in Antarctica on the basis of faunal content, Ojakangas and Matsch (1981) inferred the presence of a large marine embayment on the sites of the present Sentinel Range and Meyer Hills (Fig. 15). In this area, several hundred kilometers long, the basal Permian diamictites are not tillites, as in much of Antarctica, but dropstone conglomerates, resembling the "pebbly mudstones" of much of southeastern Asia and Asiatic Russia. Thus the region was either a very large lake (as much as 1,000 km [620 mi] long, according to Ojakangas and Matsch [1981]), or a marine embayment. Lakes this large are most uncommon and very likely would have been frozen much of the year, too much so for a thick, several-hundred-kilometer-long deposit of dropstone conglomerate to have formed. Therefore, the water body probably was marine.

Triassic. The best known fossils in Antarctica are the Triassic tetrapods. The close relations between the Early Triassic tetrapods of Antarctica and those of South Africa are well known (Colbert, 1981, 1983; Cosgriff et al., 1982; Cosgriff and Hammer, 1983). Several of the same genera, and in some cases the same species, are present in both Antarctica and Africa (Colbert, 1983). This fact does not prove that the continents were joined, but it certainly is a strong possibility. However, the connection need not have been that of two contiguous landmasses, but via a land bridge. Meyerhoff and colleagues have accumulated evidence to show that much of the Scotia Ridge, from Burdwood Bank to South Georgia along the northern flank of the Scotia Sea (Fig. 2), may once have been linked to Antarctica, and that this ridge reached its present position between Middle Triassic and Late Jurassic times (Fig. 2) (I. Taner, oral communication, 1995). It is true that several well-known Triassic reptiles described from Africa, Asia, Europe, and Antarctica have not been reported from South America, but this could easily be a collecting artifact (S. Chatterjee, oral communication, April 22, 1988).

Cosgriff and Hammer (1983) noted that four amphibian families of Early Triassic age are found together in the same bed in only three places—Antarctica, South Africa, and Tasmania (Figs. 1, 15). All four families are found in places as far away as Asia and Svalbard, but nowhere else are they found together.

Recently Hammer et al. (1990) described a new Triassic vertebrate fauna from Antarctica. This new fauna is younger than the previously discovered faunas, all of which are from the Early Triassic *Lystrosaurus* Zone. The newly discovered fauna comes from the Early to Middle Triassic *Kannemeyeria* (*Cynognathus*) Zone (Hammer, 1989). In overall composition, the newly described assemblage resembles Triassic vertebrate collections from Australia where large numbers of amphibians, especially temnospondyl amphibians, are present. In most places they outnumber the reptile taxa.

The presence of such a large temnospondyl amphibian population is believed by Thulborn (1986) to have great paleobiogeographical significance. He noted that, in Australia, the reptile-amphibian fauna is essentially endemic. This fact and the very large numbers of amphibian taxa indicated to Thulborn that, during Triassic time, Australia was largely isolated. The same arguments apply to *Kannemeyeria* Zone time in Antarctica. Not only are temnospondyl amphibians unusually abundant, but the taxa are endemic (Hammer et al., 1990); none of the taxa, even at the genus level, are known outside of Antarctica. This, however, could change as more collections from coeval strata are made.

Triassic marine strata are known now from northwestern Graham Land on the Antarctic Peninsula, and from Alexander Island, just west of the peninsula (Figs. 11, 15) (Thomson, 1975; Edwards, 1982). The faunas show close affinities with coeval

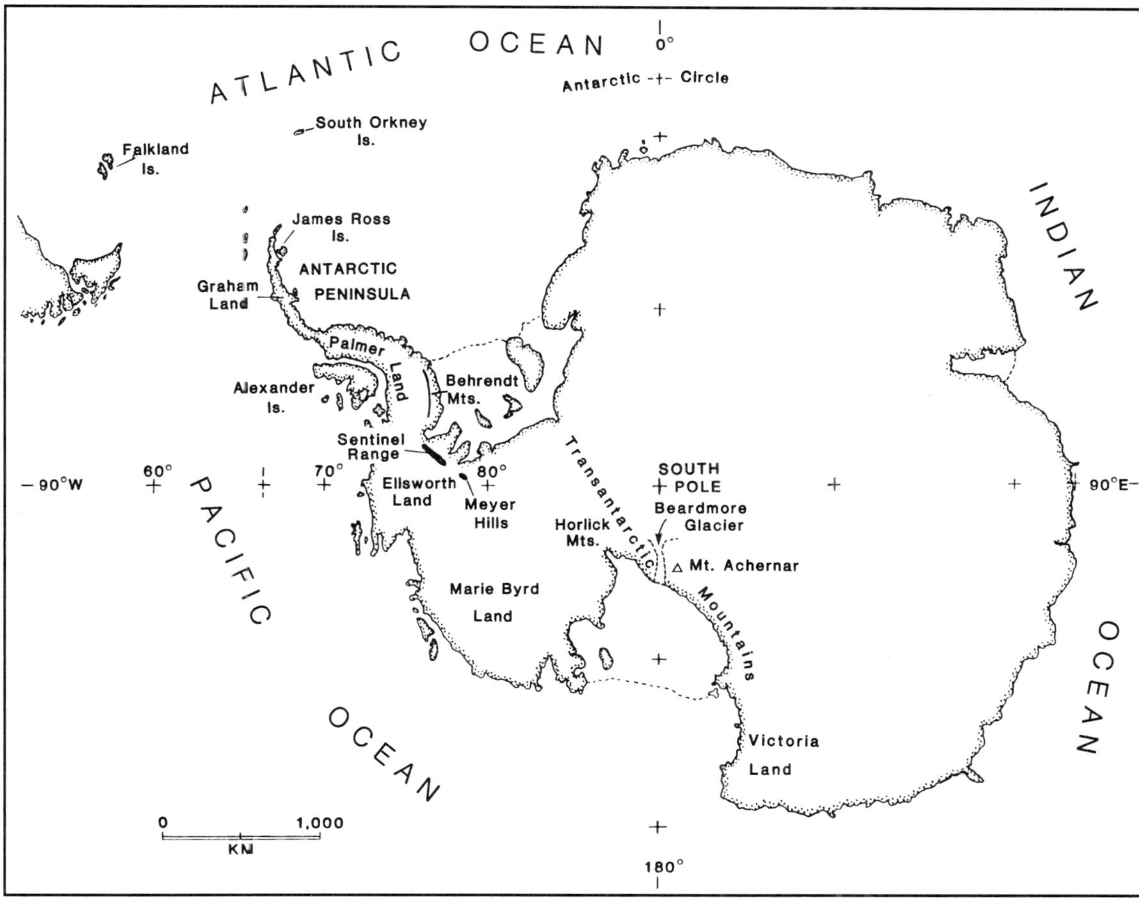

Figure 15. Index map, Antarctica, to localities discussed in the text.

faunas from New Zealand, New Guinea, and Japan (i.e., circum-Pacific faunas).

Mixed Late Triassic Tethyan and Gondwana palynomorphs recently were discovered in coastal East Antarctica. These palynomorphs are in the same layers as well-dated Late Triassic marine spinose acritarchs and a marine Triassic dinocyst according to (Foster et al., 1994, who summarized the large and growing body of evidence to show that a marine seaway occupied the present site of the Indian Ocean during at least part of Late Triassic time.

Jurassic-Cretaceous. Quilty (1970, 1977, 1982) described Middle and Late Jurassic ammonoid and bivalve faunas from Ellsworth Land (Figs. 1, 2, 12a,b, 15). These faunas are closely related to equivalent forms in Cuba, Alaska, British Columbia, Western Europe, and New Zealand. Quilty (1970) suggested that the taxa originated in Europe and dispersed to the Pacific via the Caribbean. Thomson (1981) described Middle Jurassic (Bajocian) Tethyan faunas from the south flank of the inland Behrendt Mountains (Fig. 15). Thomson (1983) also studied a Late Jurassic marine fauna from Palmer Land, and found that it is closely related to faunas in Argentina, Kenya, Kutch, Madagascar, Spiti, and the Salt Range (Figs. 1, 2, 9, 15). This distribution suggests at once the existence of a migration route from the Himalayas to the Antarctic Peninsula through the region of the present Indian Ocean. Aguirre-Urreta et al. (1990) described a Late Jurassic decapod crustacean of the Family Polychelidae from James Ross Island, just east of the Antarctic Peninsula (Fig. 15). Before, the discovery of Aguirre-Urreta et al. representatives of the family Polychelidae had been known only from Western Europe. In yet another discovery, Richter and Thomson (1989) described a marine Late Jurassic fish from a locality 25 km (15 mi) southwest of the decapod crustacean locality. The fish, a teleost, is the first record of the genus outside of Western Europe.

Thomson (1972) also described Late Jurassic ammonoids from Alexander Island and, depending on the time interval, he noted the close relations between the Antarctic faunas and equivalent ones in Argentina, Bolivia, the United Kingdom, Eastern Europe, the Salt Range, Australia, Madagascar, and California (Figs. 1, 2). Apparently the oceans had taken on a relatively modern configuration by Middle Jurassic time. Indeed, Khudoley (1974, 1988) and Khudoley and Prosorovskaya (1985) have shown that the ammonoid distributions of the entire Mesozoic are most easily and sensibly explained by dispersals through the oceans located in their present positions. Holdsworth and Nell (1992) described mixed Tethyan and high-latitude Kimmeridgian through Albian radiolarians from

Alexander Island. Hammer and Hickerson (1992) reported on Jurassic dinosaurs with varied morphological features that link them to those known from other continents.

South America

Pre-Carboniferous. Boucot and Gray (1979) summarized the evidence indicating that a north-south boundary of great importance occupied the present Andean region (Figs. 4 through 8). This boundary, which extends the full length of the present Andes, separates warm-water biotas on the west from cool-water biotas on the east. It first appeared in Cambrian time and persisted into the Permian. The boundary does shift back and forth (east and west) with time, but overall remains relatively fixed. Even during times when cool-water biotas reached the present Pacific coast, especially in Peru–Bolivia–northern Chile, the boundary simply shifted into the adjacent, present Pacific basin to reappear farther south in Chile (Boucot et al., 1980). If the same plate tectonic criteria used in the Himalaya-Xizang (Tibet) Plateau region were applied here, a 7,500-km-long (4,650 mi) suture zone would be postulated in the western part of South America 100 to 300 km (62 to 186 mi) inland from the Pacific Ocean. A careful review of the paleogeographical conditions during Devonian time as they pertain to the Andean region, indeed, to all of South America, was published by Barrett and Isaacson (1988).

In the continental areas east of the present Andes, conditions are more complex (Figs. 4–8). In the Parana basin and the adjacent Brazilian shield (Figs. 2, 5), Ordovician fossils are unrecognized, except for some Ashgillian chitinozoans recently found in the Amazon basin, but unpublished, by Yngve Grahn while at PETROBRAS, the national oil company of Brazil. However, Malvinokaffric Silurian succeeded by Malvinokaffric Devonian is widespread (Boucot, 1975; Isaacson and Sablock, 1988; Melo, 1988), and is succeeded in turn by extra-Malvinokaffric later Devonian (Boucot, 1988). In the Maranhao-Parnaiba and Amazon basins farther north, warm-water and cool-water faunas are interlayered and/or are intermixed within the same beds (Figs. 2, 5) (Boucot, 1975, 1988, 1990). In the Merida Andes southeast of Lake Maracaibo, Venezuela (Fig. 2), Malvinokaffric Realm Ordovician is characteristic (as it is in North Africa, much of central and southern Europe, North Africa and Arabia) (Fig. 5), but is succeeded by the northern North Silurian Realm (warm water: Boucot, 1975, 1988, 1990) in Europe for both the Silurian and Devonian, and in North Africa plus Arabia for the Devonian (see Figs. 5, 6). Eastern Americas Realm Devonian (also warm water) is present in the Sierra de Perija, a branch of the Andes straddling the Colombian-Venezuelan border (Figs. 2, 7a,b). The same warm-water facies extended during Silurian time to the present Atlantic coast in southern Argentina, especially south of 39°S lat (Fig. 6), where it is far south of the coeval Malvinokaffric Realm faunas of west-central Argentina (Sanchez et al., 1991). This is yet another fact that is difficult to accommodate in any tectonic model yet proposed (Cortés et al., 1984).

The belt along which warm-water and Gondwanan biotas are intercalated is similar to that which is present in Eurafrica, Asia, and elsewhere. The intercalary zone here ranges in width from a few hundred kilometers (eastern Peru, western Bolivia) to more than 1,100 km (680 mi) (Maranhao-Parnaiba, Amazon basins; Fig. 2). Much smaller distances are involved along the western rim of the continent in Chile where the transition from Pacific, northern-type faunas to Gondwanan taxa takes place in a belt only 100 to 300 km (62 to 186 mi) wide (Boucot et al., 1980; Isaacson and Sablock, 1988). By Jurassic time, practically all vestiges of Gondwana had disappeared.

Mississippian. Mississippian faunas and floras generally are relatively cosmopolitan, or Tethyan, the two terms being practically synonymous for this time. For example, the microfloras of the Maranhao-Parnaiba basin (Fig. 2; Longa, Poti Formations), the Amazon basin (Faro Formation), and southern Peru, Bolivia, and northwestern Argentina (Ambo Formation and equivalents) are very similar (Guimaraes, 1964; Petri and Fulfaro, 1983; Rocha-Campos and Archangelsky, 1985).

In northern Chile a Mississippian North American Midcontinent–type marine fauna overlies Devonian with a mixed fauna (Fig. 2; Bahlburg et al., 1987). The Devonian (Emsian-Eifelian; Fig. 3) is a mixture of warm-water *Tropidoleptus* and less abundant cool (southern) *Australocoelia*, a typical Malvinokaffric Realm taxon. The North American fauna extends even farther south, to and beyond Isla Chiloe (43°S lat; Fig. 2) to 55°S where Pennsylvanian, warm-water, fusulinid-bearing limestone is present (Cecioni, 1955; Gerth, 1957; Douglass and Nestell, 1976).

According to Harrington (1962) and Gonzalez (1990), mountain glaciation began during Mississippian time in southern Bolivia and in western parts of Argentina. Gonzalez (1990) dated the glaciation as late Visean to early Namurian (late Visean–early Serpukhovian; Fig. 3). Lopez-Gamundi et al. (1993) discussed Namurian-Westphalian diamictite from the Paganzo Basin (30°S–68°W).

Pennsylvanian. Well-differentiated faunal and floral regions were present by Pennsylvanian time. In the Maranhao-Parnaiba basin, for example, the Piaui Formation has a Gondwana microflora and a Tethys (North American Midcontinent) invertebrate marine fauna (Fig. 2). In the Amazon basin, the Pennsylvanian Monte Alegre and Itaituba Formations have a North American Midcontinent marine fauna, a Gondwana microflora, and a North American megaflora (Derby, 1894; Guimaraes, 1964; Petri and Fulfaro, 1983; Rocha-Campos and Archangelsky, 1985). In southeastern Peru and western Bolivia, however, the microflora is Gondwanan, the marine invertebrates are North American Midcontinent (with many endemics related to the Amazon basin), and the megaflora is ecotonal—definite Gondwana forms mixed in the same strata with northern taxa (Newell et al., 1953; Oviedo, 1965; Castaños and Rodrigo-G., 1978; Rocha-Campos and Archangelsky, 1985). Rocha-Campos (1972) has noted that the Amazon and Peru-Bolivia faunas are a distinctive province, having taxa endemic to the region as well as North American Midcontinent–Andean taxa. This is in sharp contrast to the biotas of the Parana basin and northwestern Argentina, which are truly Gondwanan (Harrington, 1962; Rocha-Campos, 1979; Rocha-Campos and Archangelsky, 1985). The area of mixing, however, is a large one, whether the mixing is by intercalation or by mixing in the same stratum. Several mixed North American and Gondwanan flora localities have been described in the Parana basin of Brazil and in southern Argentina (Archangelsky, 1960; Archangelsky and de la Sota, 1960; Lacey, 1975; Vozenin-Serra, 1984).

We have mentioned the warm-water northern Pennsylvanian south of latitude 43° in southern Chile.

Permian (Figs. 8, 10). The shallow-water, carbonate-dominated Copacabana Group of southeastern Peru and western Bolivia is one of the best-known South American Pennsylvanian-Permian sequences (Fig. 2). In the Lake Titicaca region, the Pennsylvanian-Permian northern faunas overlie an Ordovician, Silurian, and earlier Devonian sequence dominated by cold-water Malvinokaffric Realm biotas (Boucot and Gray, 1979). These Malvinokaffric Realm strata extend far north and south of the Lake Titicaca region.

The lower part of the Copacabana Group in the Titicaca region belongs to the North American Midcontinent–Andean Realm with many forms congeneric and conspecific with northern taxa (Dunbar and Newell, 1946; Newell, 1949; Newell et al., 1953; Ahlfeld and Branisa, 1960; Chamot, 1965; Cousminer, 1965; Oviedo, 1965; Rocha-Campos, 1973, 1979; Teichert, 1974; Castaños and Rodrigo, 1978; J. R. P. Ross, 1979; Dickins, 1985b; Wilson, 1990). Its equivalent extends across eastern Peru into the Amazon and Maranhao-Parnaiba basins (Guimaraes, 1964; Petri and Fulfaro, 1983).

The upper part of the Copacabana Group is Gondwanan (Cousminer, 1965). Thus the southeastern Peruvian–western Bolivian Pennsylvanian–Permian sequence is an excellent example of the intercalary relations through time between Gondwana and warm-water conditions; in this area there are at least two intercalations of each. Cousminer (1965), incidentally, found that the microflora of the upper Copacabana is closely related to equivalent taxa in China, in addition to taxa in India and Australia.

The equivalent section in the Maranhao-Parnaiba basin (Pedra da Fogo Formation) is largely Gondwanan. However, in addition to the Gondwanan plants listed by Guimaraes (1964), Dolianiti (1972) found Boreal genera. Petri and Fulfaro (1983) also mentioned that a Lower Permian amphibian in this basin has its closest relatives in the Urals.

Glaciation during Early Permian time took place mainly in the Parana basin region, eastern parts of Argentina, and in the Malvinas (Falkland) Islands (Harrington, 1962; Gonzalez, 1990).

The influence of the warm-water Copacabana invasion of South America extended along the Chilean coast to 55°S lat. In the Parana basin of Argentina, the Lower Permian has a mixed Boreal and Gondwanan flora, whereas the Upper Permian is Gondwanan (Rocha-Campos, 1973).

Triassic. Rocha-Campos (1973) observed that the Triassic distribution of biotal realms was similar to that in the Permian—mainly Andean (Tethyan, North American) north of 20°S lat, and Gondwanan south of that latitude (Fig. 11). The Triassic fauna studied by Kummel (1950) in northern Peru is typically Tethyan. However, Triassic seas invaded South America in very few places (Harrington, 1962). De Oliveira (1956), writing of the freshwater and terrestrial biotas of Brazil, reported a unionid from Sousa in Paraiba State, northeastern Brazil. J. B. Reeside, Jr., who studied this specimen, stated that it is closely related to unionids in the North American Triassic (de Oliveira, 1956).

The floras show the same separation of realms, but are very inadequately studied. As for the vertebrate faunas, these are moderately well known (Triassic especially), and have been covered in the preceding sections. Keyser (1981), writing of the Late Permian and Triassic tetrapods of Africa, said that the Upper Permian taxon *Endothiodon* occurs also in Brazil and India. As for the Triassic, he commented (p. 63), "the Dicynodontia of the Upper Triassic of South America are not known from Africa, although there is some resemblance with the forms from the northern hemisphere." Thus, the tetrapods, as elsewhere (in and out of Gondwana) show close connections among some of the continents, especially in the Northern Hemisphere; whereas the marine invertebrates show the omnipresence of seaways throughout the postulated Pangaea.

Eastern North American Triassic-Jurassic Basins

A landmark paper by Sues and Olsen (1990) has revealed the presence in the eastern North American Newark Supergroup of Late Triassic–Early Jurassic Gondwanan vertebrates (Fig. 11). The Triassic locality is in the Richmond basin, 19 km west of Richmond, Virginia (Fig. 2). Here in beds of early Late Triassic age Sues and Olsen (1990, p. 1020) found:

> ". . . abundant remains of a diversified assemblage of small- to medium-sized tetrapods that closely resembles Southern Hemisphere (Gondwanan) assemblages in the predominance of certain synapsids (mammal-like reptiles). Associated palynomorphs indicate an early middle Carnian age for the fossiliferous strata. The discovery suggests that previously recognized differences between tetrapod assemblages of early Late Triassic age from Gondwana and Laurasia at least in part reflect differences in stratigraphic age, rather than geographic separation".

Yet had this discovery been made in a remote region that is little studied, such as Tibet, a suture zone would likely be postulated by some plate tectonicists somewhere within or just west of the Newark Supergroup basins!

Shortly after the Sues and Olsen (1990) paper appeared, Shubin et al. (1991) reported the presence in Nova Scotia's Fundy basin of an Early Jurassic taxon, the cynodont *Pachygenelus*, known previously only from the upper Stormberg Group of South Africa. Thus the Sues and Olsen (1990) and

Shubin et al. (1991) discoveries prove the existence of yet another major intercalary zone, in this case, a belt of continental sediments in eastern North America reaching to at least 46°N lat and containing, in vertical succession, at least two intercalations each of Gondwanan and northern tetrapod faunas. This particular example is most instructive because it comes from a group of rocks that have been mapped and studied for more than 150 yr and, through sedimentary provenance studies, proved to be an integral part of North America (e.g., King, 1951; Parker et al., 1988). (As a comparison, in plate tectonic logic, the Cambrian, Early Ordovician, Silurian, Early Devonian, and Mississippian of Nova Scotia, parts of Newfoundland, southeastern Maine, southeastern New Brunswick, eastern Massachusetts, and Rhode Island are of European aspect. Hence the possibility of migrating plates, in a biogeographical sense, exists for these earlier Paleozoic intervals. Beginning with the Pennsylvanian, however, it is not possible to shift around parts of eastern North America. This is true because of the presence in both Europe and eastern North American of a relatively uniform Euramerian flora, accompanied by a relatively uniform tetrapod biota.)

PANGAEA AND THE LATE CRETACEOUS

We have given many examples of the use of paleobiogeographical data in paleogeographical reconstructions. Our principal theme has been that the known distributions of fossil organisms lend themselves more readily to an earth model like that of today than to a Pangaeic model (cf. Fig. 7a with 7b and Fig. 12a with 12b; see figure captions for more detail). Therefore, it is interesting to examine a case involving an interval of time that all earth scientists would agree must be resolved using approximately the present distribution of continents and ocean basins. This case involves a collection of continental Late Cretaceous vertebrate fossils from south of Vitoria in the Basque country of north-central Spain (Fig. 1). Marine Campanian strata underlie the vertebrate-bearing beds; strata with Maastrichtian shark teeth, microfossils, and the diagnostic Maastrichtian rayfish, *Rhombous binkhorsti*, directly overlie the vertebrate-bearing beds. Thus the fossil collections are assumed to be early Maastrichtian; they are not older than the Campanian (Cappetta, 1987; Astibia et al., 1990).

The collecting locality is a sand quarry near Lano, approximately 20 km (12 mi) southeast of Vitoria. Several thousand bones and teeth have been extracted from the quarry since 1987. Astibia et al. (1990) have reported not less than 25 families identified to date, which include many representatives of fish, amphibians, and reptiles; as well as a few mammals. Of the taxa identified to date, several abelisaurids and a large collection of titanosaurids are present. Although a few isolated specimens are known from the Late Cretaceous elsewhere in Western Europe, the exceptional numbers of these two families at this locality are normally found only in Gondwanan areas. Of particular importance is the discovery of a madtsoid snake, *Madtsia* sp., known previously only from the Late Cretaceous of Niger, Madagascar, Argentina, and Brazil, and from the Cenozoic of Morocco, Argentina, Brazil, and Australia. It is the first occurrence of the snake from a Laurasian area. It is not difficult to imagine some of the paleogeographical explanations that would have been conjured had this vertebrate find involved Jurassic or older taxa!

INTERCALARY ZONE

Figures 16 through 18 summarize the data presented in the preceding sections. Figure 16 shows, by age, the occurrences of northern biotas in the southern sphere. The figure also shows by area, the occurrences of southern biotas extending into the northern sphere. Such a distribution clearly requires many large, open water bodies permitting north-south communication for the marine faunas, and land (including closely spaced islands) connections for certain tetrapods and plants.

Figure 17 shows (with diagonal shading) the overall width of the intercalary zone for the entire time interval considered here, Cambrian through Early Cretaceous. For comparison, the northern boundary of Early Permian Gondwana-type fossils is superimposed. Just as Sues and Olsen (1990) and Shubin et al. (1991) concluded that the reptilian occurrences, Gondwanan and northern alike, in the Late Triassic–Early Jurassic of eastern North America reflect differences in stratigraphical age, we likewise conclude that some of the northern-southern overlaps illustrated in Figure 17 reflect differences in stratigraphical age. In many areas, however, the northern and southern overlaps are the result of physical mixing, a phenomenon that can be observed in very many places.

In Figure 18 we have plotted the northernmost and southernmost extents of selected biotas. These include (1) the northernmost extent of Malvinokaffric Realm marine invertebrates during pre-Hirnantian Ordovician time; (2) the northernmost occurrences of Triassic tetrapods of Gondwanan origin (*Lystrosaurus* and younger zones); (3) an outline of the maximum extent of the Gondwana Realm during Early Permian time; and (4) the southernmost known penetrations of Triassic warm-water (Tethyan or northern) marine invertebrate faunas.

The figures show the exceptional pervasiveness of marine access in the Southern Hemisphere during the times of the Malvinokaffric and Gondwana Realms. The locations of the continental and mountain glaciers of Late Ordovician and Early Permian times demonstrate that many of these seaways had to be deep-water, ocean-type bodies (Brooks, 1926, 1949). The two figures also show that the overall motions of biotas were north and south, parallel with modern climatic zone expansions and contractions (as during the Pleistocene glaciations), and in the same directions as today's overall biotal movements. These figures therefore provide strong evidence for the concept that the general positions of the continents and ocean basins have not changed significantly during Phanerozoic time. Had we used a Pangaeic reconstruction of the Earth for these base maps, the glaciated areas would have occupied (in part) remote areas of interior Pangaea where moist, warm, oceanic air could not have

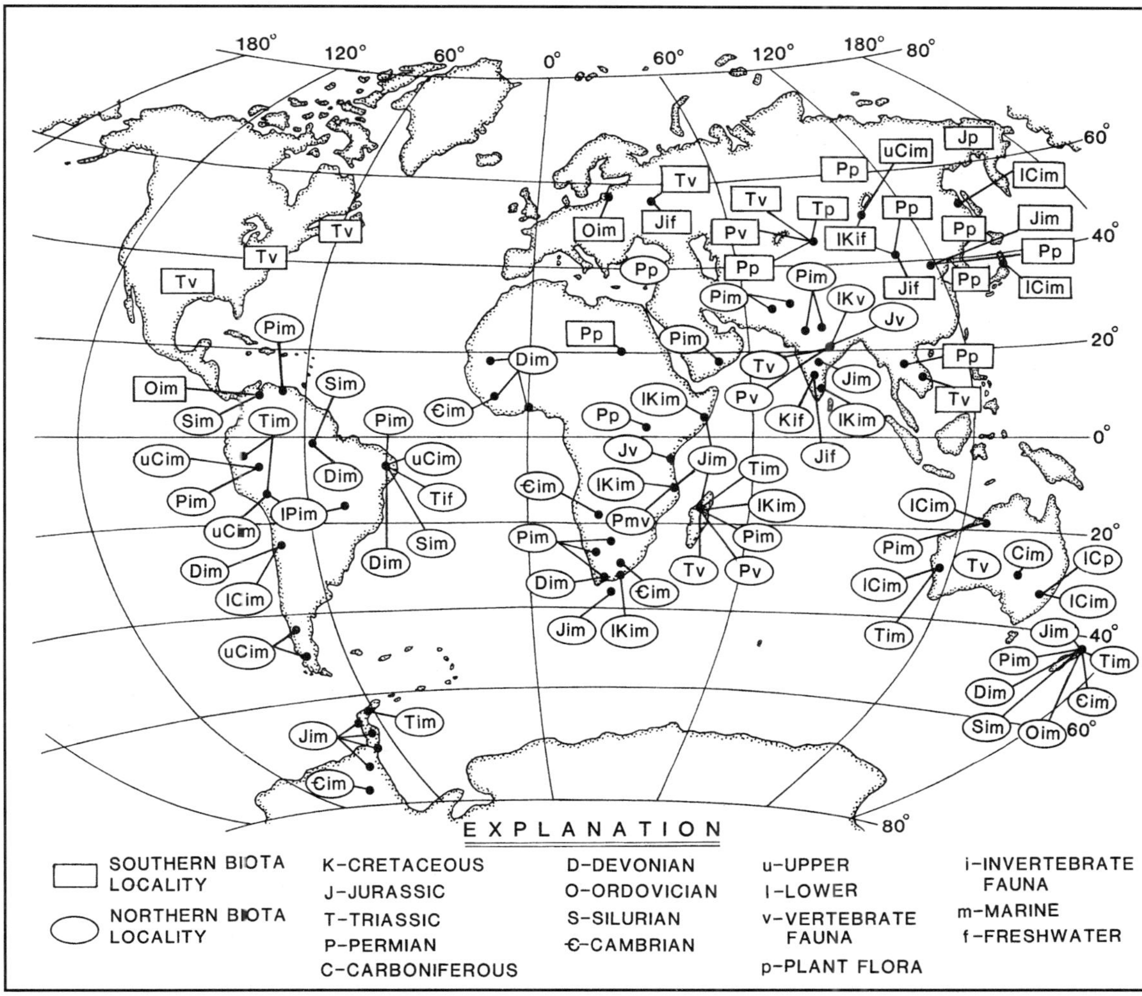

Figure 16. Map showing southern fossil localities in the Northern Hemisphere (rectangles) and northern fossil localities in the Southern Hemisphere (ovals). See text for discussion of the individual localities.

reached, and the biotal migrations (and therefore the climatic variations) would have been unsystematic and at times random. The data reviewed here, therefore, are more consistent with the present distribution of continents and ocean basins than with a Pangaeic distribution. Finally, the figures also demonstrate that the known faunal and floral realms of the past do not coincide with the various plates that are postulated in plate tectonics.

FACTORS RELATED TO THE PANGAEA PROBLEM

Evaporites

Evaporites commonly are thought to form in warm, or hot, climates where water evaporation exceeds water influx. Meyerhoff (1970a,b) found that 95% of all of evaporites from the Late Proterozoic to the present, by volume and by area, lie in regions which today receive less than 1,000 mm (40 in) of rainfall. At least 35% of these evaporites are pre-Permian; a rather remarkable statistic and powerful argument for stable continental cratons through Phanerozoic time (Lowman, 1985, 1986), because thick evaporite sequences form on stable cratonic blocks and (not as a rule) in highly mobile belts (Zharkov, 1981). However, very good geochemical evidence has now been accumulated by Hardie (1990, 1991) to show that factors other than climate are also extremely important in evaporite formation.

Late Paleozoic evaporites are associated with areas where reefs and fusulinids thrived. They were deposited generally adjacent to areas where warm oceanic currents were present, even at high latitudes, in Late Proterozoic and early Paleozoic times (Meyerhoff, 1970a,b).

All evaporites in the past, except for the occasional times when high-latitude deposits could form, lie in a belt axisymmetrical with, and tilted at about 23° to, the present equator; this includes the evaporite accumulations of today (Teichert, 1964). This fact, plus the reasons discussed in the preceding two paragraphs, suggests that the present geography of the continents and ocean basins approximates the geographies of the continents and ocean basins in the Phanerozoic past. Strongly supporting this conclusion is the fact that faunal and floral dispositions of

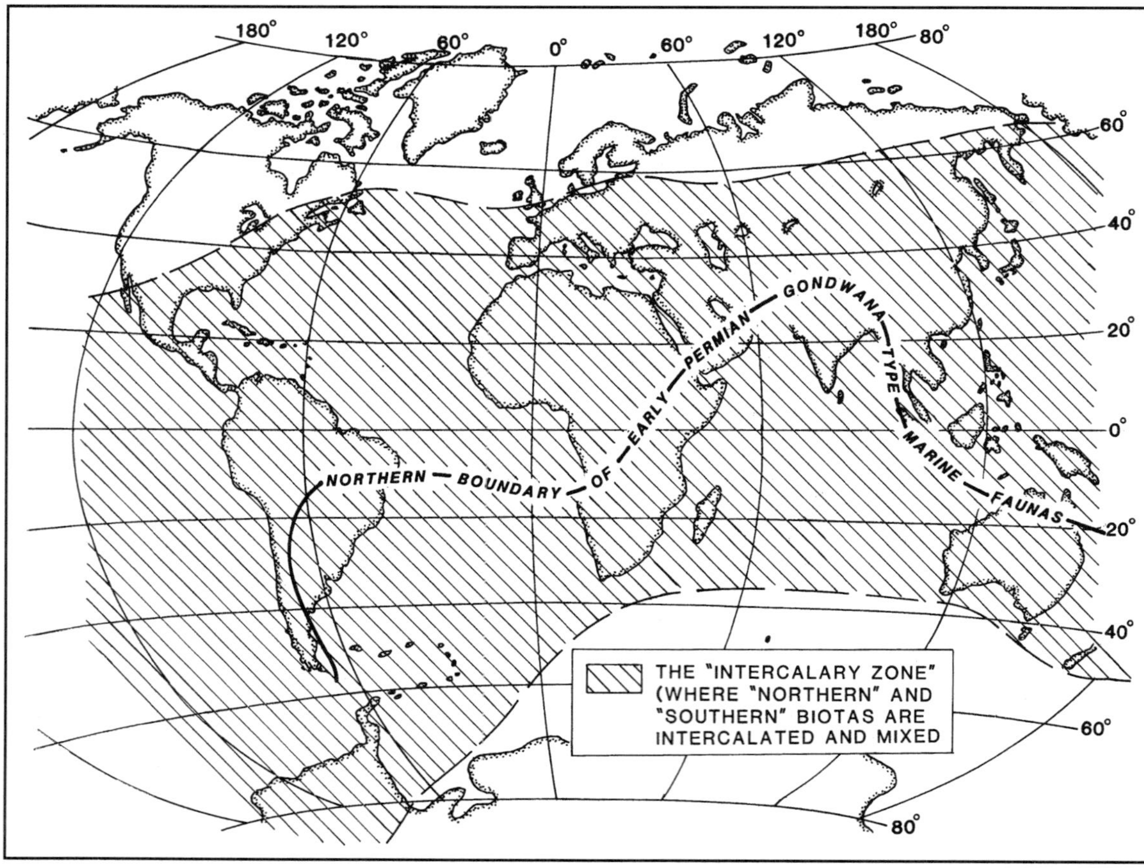

Figure 17. Map showing the "intercalary zone" of Phanerozoic time (diagonal shading). This is the zone where the northern and southern biotas are intercalated or mixed in the same bed. The map is based on the data in Figure 16. The heavy line marks the northern known limit of the early Early Permian marine fossil localities. The northern limit of the intercalary zone is the northern known limit of southern taxa of all ages and the southern limit is the southern known limit of northern taxa of all ages. See Figures 16 and 18.

the post-Devonian are bipolar to the present globe. The distribution of the evaporites is closely related to the distribution of certain faunal groups (such as reef-building organisms, fusulinids, and miliolids) and therefore does not fit any currently popular plate tectonic reassemblies. For plate tectonics to succeed, both the biotas of the past and the climatic indicators preserved in the stratigraphic record (such as evaporites) must be used in conjunction with geophysical and structural geological data to find an acceptable plate reassembly.

Coals

Many of the statements made for evaporites apply also to coals, which comprise one of the largest collections of fossil biota in existence. During most times after the Devonian, two axisymmetrical globe-encircling belts of coal deposits were present (Meyerhoff and Teichert, 1971); one lies north of the tilted evaporite belt, the other lies south of it. A third coal belt, this one tropical, existed at times, especially during the Cenozoic. This disposition clearly suggests once again that the present geography of continents and ocean basins resembles that of the past. Regardless, coals—because they are an integral part of paleobiogeography—also must be accommodated within any tectonic model if that model is to be valid.

Changing widths of climatic zones

As we pointed out in our discussion of evaporites above, the evidence for changing widths of the Earth's climatic zones is clear and unequivocal (Brooks, 1926, 1949; Wegener, 1929; Flint, 1947; Florin, 1963; Mercer, 1983; Funder et al., 1985; Francis and McMillan, 1987; Kauffman, 1987; Paul, 1988; Burckle and Pokras, 1991; Taylor et al., 1992). Despite this evidence, Drewry et al. (1974), Habicht (1979), Hallam (1989), and many others who have built a Pangaea on the basis of postulates and data seem to attach greater importance to the postulates than to the data. One of these postulates is that the widths of climatic zones have remained essentially constant throughout Phanerozoic time.

If the widths of the climatic zones have remained unchanged, how else to explain the presence of large Cretaceous dinosaurs and trees in such high-latitude localities as Svalbard

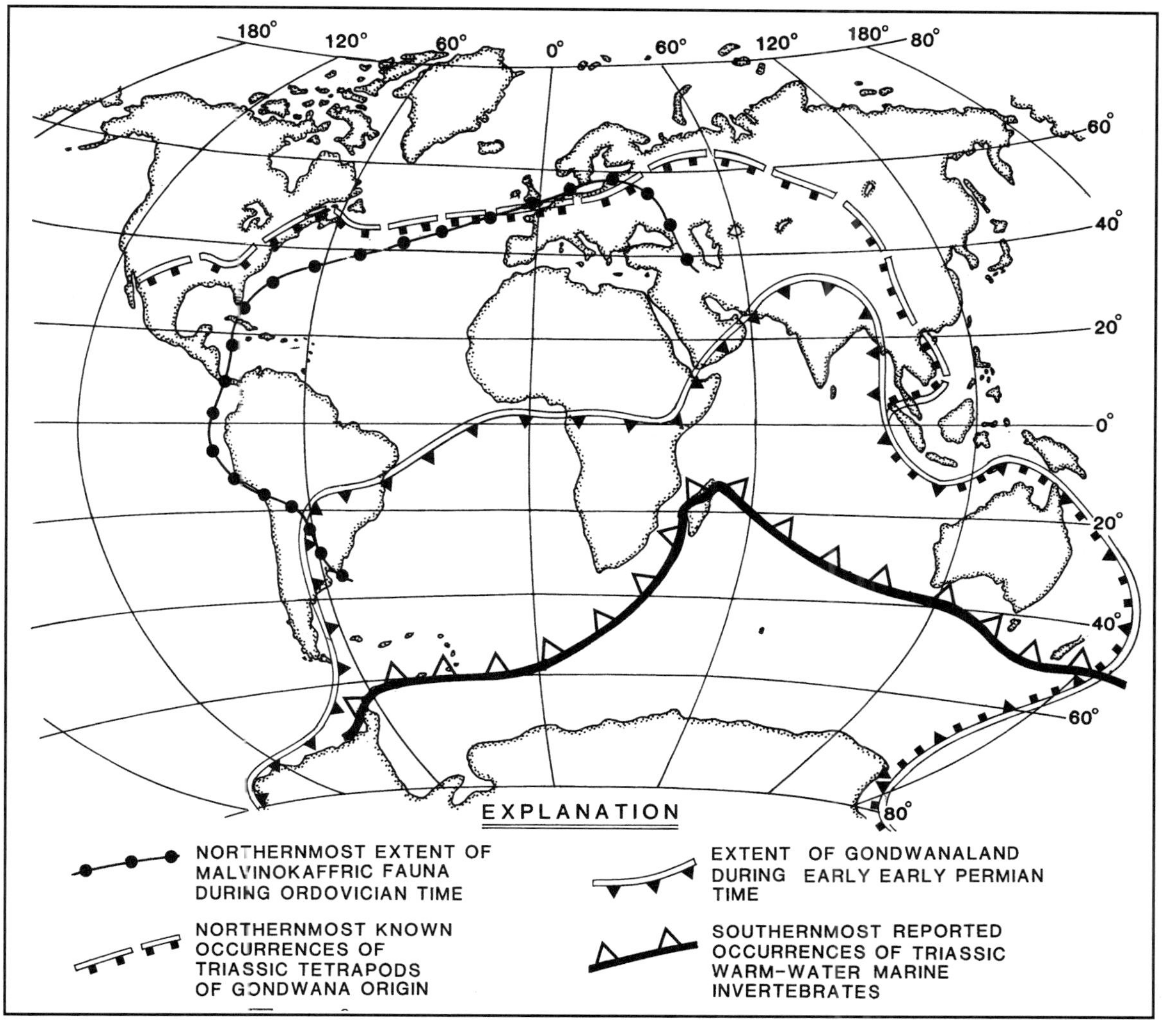

Figure 18. Map showing (1) the northern limit of Malvinokaffric faunas during Ordovician time (based in part on Boucot and Gray, 1983); (2) the northernmost known occurrences of Gondwanan Triassic tetrapods; (3) the outline of Early Permian "Gondwanaland"; and (4) the southernmost reported occurrences of Triassic temperate and warm-water marine invertebrates. This map and Figure 17 demonstrate that the faunal and floral distributions of Middle Cambrian–Early Cretaceous times are nearly identical with those of Cenozoic time as demonstrated also by Florin (1963). The northern and southern limits of the intercalary zone, as well as the northern and southern limits of specific groups of organisms as shown here, demonstrate a distinct bipolarity in the distribution of Middle Cambrian–Early Cretaceous organisms and are explained most simply as a consequence of north-south migrations on a modern globe.

(Fig. 1) and the present North Slope of Alaska (Fig. 2) (de Lapparent, 1962; Paul, 1988)? What other explanations can be found for late Paleocene–middle Eocene forests on Ellesmereland with crocodilian bones, palm trees in west-central Greenland and southern Alaska (Hollick, 1936), nummulitic (Tethyan) limestone on the Hatton-Rockall Plateau, and mangrove swamps in the London-Paris basin (Figs. 1, 2) (Florin, 1963; Francis and McMillan, 1987; and many other references)? Since early Pliocene time alone, the width of the temperate zone has changed more than 15° in both the Northern (Funder et al., 1985) and Southern Hemispheres (Burckle and Pokras, 1991), a total distance in each hemisphere of 1,650 km (1,025 mi)!

For the latest Cretaceous and Cenozoic items mentioned here, one cannot appeal to a plate tectonic explanation but must accept the fact that the climatic belts had widths that were very different from those of today. For Paleozoic time, by using a plate tectonic explanation one can try to rationalize the Mississippian evaporites of eastern Canada that are overlain by coal-bearing Pennsylvanian strata (a situation also common in western Europe), but the alternative of changing widths of the climatic zones must also be considered. Admittedly it is more difficult to be certain about Paleozoic events than of Cenozoic events, yet the simplest explanation (in this case, a nonmechanical one without subduction zones, continental sutures, and so forth) is far more likely to be the correct one as experience in geology has demonstrated repeatedly (e.g., Dallmus, 1958). Thus, as Dickins (1985b,c) pointed out for the Carboniferous, Permian, and Triassic, concerns about the necessity for having a broad Tethyan climatic zone—broader than its present equivalent, the torrid zone—are without foundation (e.g., Gobbett, 1967, 1973a; Hill, 1973; Minato and Kato, 1977; Spörli and Gregory, 1981; and many others). The accuracy of this statement is highlighted by the fact that, in any worldwide Pangaeic reassembly, some major evaporite deposits have to be placed at very high latitude (Meyerhoff and Teichert, 1971). And, if any proof of great climatic latitude change were needed for Permian time, one must only consider the significance of the Permian bauxite group mineral boehmite occurring in New South Wales (Loughnan, 1975), a region that elsewhere includes Permian stratigraphic sections with glacigene rocks! Boehmite is known to form only under relatively warm and humid conditions; i.e., it is incompatible with a cold, glacial climate. The New South Wales Permian boehmite indicates the presence there of at least one brief interval of warm-humid climate in an otherwise cool to cold climatic regime.

Early Permian stream base level

During Early Permian time, extensive mountain glaciation took place in South America, Africa, India, Australia, and possibly Antarctica; lesser glaciation took place in the Polar Urals and northeastern Asia (Meyerhoff and Teichert, 1971). A large number of these valley glaciers emptied into the sea towards the present coastlines as proved by the extensive intertonguing of Early Permian marine and glacial deposits in such widely scattered areas as Brazil (Harrington, 1962), southern Africa (Martin, 1953, 1975; du Toit, 1954; Haughton, 1963, 1969), East Africa (Kent et al., 1971), Madagascar (Besairie, 1972), and Australia (Teichert, 1951, 1958; McWhae et al., 1958; Bowen, 1964; Meyerhoff and Teichert, 1971). Early Permian valleys are being reexcavated today by modern streams in scores of places. In Zaire, for example, the principal modern drainage in the mountainous eastern part of the country was also the Permian drainage (Boutakoff, 1940, 1948; Meyerhoff and Teichert, 1971). In Namibia, Martin (e.g., 1953, 1975) proved that the streams entered a basin on the site of the present South Atlantic Ocean. The same can be shown in Tanzania, where drainage was into a basin on the site of the present Indian Ocean (Kent et al., 1971), and in Australia, where drainage was into basins on the sites of the present Pacific and Southern Oceans (Bowen, 1964). These facts indicate that base levels for the various southern continents during Permian time were marine basins located where the South Atlantic, Southern, and Indian Oceans lie today.

Gondwana's erosion products

If Gondwana's Early Permian streams drained into basins now submerged offshore, another problem of Gondwanan geology may be explained. This is the fact that the volume of Gondwana's erosion products in African onshore basins (e.g., Great Karoo and Kalahari basins) is too small to explain the volume of pre-Karroo rock that has been removed (Martin, 1975). Clearly no precise figures can be determined. Yet order-of-magnitude calculations of Pennsylvania–Early Cretaceous sedimentary volumes give 1.5 to 2.0×10^6 km^3 (0.9 to 1.2×10^6 mi^3), whereas the estimated volume of pre-Karoo rocks removed is more than twice this amount (4.0×10^6 km^3; 2.4×10^6 mi^3). Nor can post–Early Cretaceous erosion account for the missing rocks (Martin, 1975). For example, boreholes drilled below the Karoo have revealed the presence of marine Cambro-Ordovician in several areas (Zaire, Etosha basin, other areas: J. A. Momper, oral communication, 1984; Daly et al., 1992), yet little or no Cambro-Ordovician is reported (Drysdall et al., 1972) from surface outcrops. However, if seaways during Paleozoic time were present where the existing ocean basins are located the problem vanishes. Martin (1975) has published a most thoughtful discussion of this vital problem.

Glaciation and the need for oceans

A powerful argument for the existence during later Ordovician and late Paleozoic times of oceanic basins within Gondwanaland is the fact that glaciation took place at both times on such a large scale. Brooks (1926, 1949), Salamon-Calvi (1933), Meyerhoff (1970a), and Meyerhoff and Teichert (1971) have pointed out that glaciers cannot cover continents the size of Pangaea for the simple reason that it is physically impossible to initiate and nourish glaciers in the interiors of such large continents. Instead, the size of Pangaea means only that its interior would have been a vast desert land like parts of interior Siberia today, and not a region of continental glaciers. Glaciation requires the interaction of warm ocean currents, moisture-laden warm air, and cold winds generated by glacial ice (Brooks, 1949).

Moisture to sustain continental glaciers cannot be carried far (2,500 km [1,550 mi] maximum; the diameter of Gondwana generally is perceived as having been much greater). Nor can epeiric seas have provided the moisture because like Hudson Bay today such seas freeze over during the winter months, thereby preventing evaporation. In fact, Salamon-Calvi (1933), once one of Wegener's staunchest supporters, recognized in his 1933 book that warm oceanic basins were required to feed the Permian glaciers.

Just as important is the fact that this same moisture must be available to nourish luxuriant growth of the type that is known in Carboniferous and younger sediments of Gondwanaland. The vast coal measures that we see would not have formed had the continents been joined; instead deserts, not swamps, would have been present. Moreover, the large tetrapod population had to have these plants and swamps for their very existence.

A related fact is that 88% of the world's economic coal deposits are on the eastern sides of the continents (North and South America, Africa, India, Australia, Asia) or in northwestern Europe and the Arctic coast of northwestern Asia (Meyerhoff and Teichert, 1971; Teichert and Meyerhoff, 1972). This phenomenon has been called the east-side rule (Meyerhoff and

Teichert, 1971). These are precisely the regions where today the heaviest amounts of rain fall in the temperate zones. The reason why northwestern Europe and the adjacent part of arctic Asia are exceptions to the east-side rule is that it is only in the North Atlantic Ocean where a major moisture-bearing wind system and warm ocean current, the Gulf Stream, cross from the eastern side of a continent (North America) to the northwestern side of another (Eurasia). If the North Atlantic existed during Permian time as now—the thick coal measures of the eastern sides of the continents and of northwestern Eurasia—are easily explicable (Meyerhoff and Teichert, 1971). Otherwise, they are a puzzle because they could not have formed in the interior of a giant moistureless supercontinent.

High-latitude life

Another theme that recurs constantly is the statement that large continental movements have taken place because big trees, widespread vegetation, and abundant, large animals could not have survived in the polar areas of long nights, when chlorophyll could not have formed and everything was too cold.

The remains of large tetrapods have been found in high latitudes in strata as old as the Devonian (Jarvik, 1961; Spjeldnaes, 1982; Brouwers et al., 1987; Paul, 1988). The examples described by Brouwers et al. (1987) and Paul (1988) involve very large Late Cretaceous tetrapods that fed on luxuriant plants. No one doubts seriously that continental movements in polar regions have been small since Late Cretaceous time. Florin's (1963) maps show large plants (and tetrapods by inference) at extremely high latitudes in Permian to Quaternary times.

The same argument has been applied to the presence of large trees. This argument has always puzzled us, because large trees live in parts of the Arctic today in a much colder climate than usually prevailed in the past. Some sizable trees in Siberia live as far north as 73°N. The discovery of extensive Tertiary forests with large-diameter trees in Arctic Canada at 80°N lat. on Ellesmere Island demonstrates unequivocally the error of this widespread belief (Francis and McMillan, 1987), as do the presence of large Pliocene trees in fossil forests at 82°30'N in northern Greenland (Fig. 2) (Funder et al., 1985) and 83°30'S in the Beardmore Glacier area of Antarctica (Fig. 16) (Burckle and Pokras, 1991), and the recent discovery of a forest of Late Permian age, interpreted to have lived between 80° and 85°S, on Mt. Achernar in the Transantarctic Mountains (Taylor et al., 1992).

CONCLUSIONS

1. Biogeographical data comprise a powerful tool for resolving geotectonic problems. Such data, however, are rarely used in existing plate tectonic models. Yet our study demonstrates that plate boundaries generated by geophysical and geological theory generally do not coincide with biogeographical boundaries based on extensive and detailed faunal and floral studies backed by field mapping. For a successful geotectonic model to be built, this conflict must be resolved and specialists in structural geology, geophysics, stratigraphy, and paleontology must work together to resolve the existing major conflicts.

2. Our study reveals that Paleozoic and Mesozoic faunas and floras are intercalated across a broad east-west zone (the intercalary zone) that in places is fully 5,000 km (3,100 mi) wide (Fig. 17), extending from western South America to the Pacific shores of Asia, Australia, and New Zealand. Such a possibility is not incorporated into any existing plate tectonic model.

3. It makes no sense to shuttle the continents in repeated to-and-fro motions to explain anomalous occurrences of Gondwana elements in northern continents and vice-versa (Mu Enzhi et al., 1986). At present, data from Asia, greater Australia, Africa, and South America suggest that the boundary between the northern faunal realms and southern (Malvinokaffric and Gondwanan) faunal realms is a broad intercalary zone where changes from one realm to another are gradual, not abrupt (see also Smith, 1988).

4. Another result of uniformly applying plate tectonic principles is to conclude that Phanerozoic suture zones equivalent to the Taurus-Zagros-Indus-Yarlung suture zone must be present somewhere beneath both the Sahara sands and the Amazonian jungles of South America. Field geological data eliminate these possibilities.

5. Further, if we apply these principles to all paleobiogeographical realm studies, we should find major sutures between the Cathaysian Realm (eastern Asia) and the Angara Realm (central Asia), between eastern and western Australia, inland from and parallel to the Pacific coast of South America, and beneath or adjacent to the Newark basins of eastern North America (a Mesozoic suture). No such sutures have been identified. Therefore, the necessity for a suture zone between Gondwana and the northern realms seems to be nil. In any case, no existing plate model fits the paleobiogeography. Hence, the models of Dietz and Holden (1970), Drewry et al. (1974), and Johnson et al. (1976) must be incorrect, and a new model that satisfies *all* the data must be constructed.

6. The data reviewed here indicate that, during the Phanerozoic, seaways were widespread at different times within both the northern and southern continents; deep water at one time or another was almost everywhere where it is today. These same data also show that free migration of tetrapod faunas from continent to continent was possible many times during geological history, although some areas (e.g., Australia–New Guinea, New Zealand, and Antarctica) seem to have been isolated during long periods of time. The data also show that India was never completely isolated from Asia during Phanerozoic time (Chatterjee and Hotton, 1986), although long periods of isolation are required by most tectonic models (e.g., Dietz and Holden, 1970; Drewry et al., 1974; Johnson et al., 1976).

7. Every marine basin in Gondwanaland (and its Malvinokaffric predecessors) either borders an existing ocean basin or opens into an existing ocean basin. This fact suggests that the present geographical relations among the various ocean basins

and continents may have changed very little during Phanerozoic (and likely earlier) time.

8. The presence of large volumes of ocean water in southern areas is required to explain every marine locality mentioned here. Even the cold-water Malvinokaffric and Gondwanan faunal realms require the presence of large expanses of ocean water. In fact, warm northern water surrounded, or partly surrounded, the Malvinokaffric and Gondwanan areas during every interval of Phanerozoic time (Figs. 4 through 8, 10 through 13, 17).

9. Limited studies of Mississippian through recent faunas and floras show a bipolar distribution of biotas much like that of the present. This too is a strong argument suggesting that the geographical relations among existing continents and ocean basins have changed little, at least since the beginning of the Mississippian.

10. The many contradictions within current tectonic models have led to the construction of new hypotheses that attempt to utilize all of the data (e.g., I. Taner, oral communication, 1995). Hence, now is the time for geophysicists, structural geologists, stratigraphers, and paleontologists to work together to see whether a satisfactory model can be worked out, a model that accounts for all, and not just part, of the data.

ACKNOWLEDGMENTS

We are grateful to J. A. Grant-Mackie, University of Auckland, for having reviewed an earlier version of the manuscript, and for having provided considerable insight into varied New Zealand questions. We thank Maurice Kamen-Kaye for suggesting the use of the term *intercalary* in the subtitle of this volume, and for his data from East Africa, Madagascar, and the Seychelles, as well as the Persian Gulf. Michael Cooper, University of Durban–Westville, very helpfully reviewed some crucial South African items. Brian Glenister, University of Iowa, kindly provided insight into some of the Permian problems. Margrit Taggart kindly supplied copies of several hard-to-obtain papers by Dr. Teichert. B. K. Tan and T. T. Khoo of the University of Malaya supplied numerous papers on the Paleozoic of Southeast Asia. We thank J. David Love and Charles J. Smiley for excellent, thorough constructive reviews; a third, unidentified critic supplied useful suggestions. Finally, we thank Kathryn L. Meyerhoff for drawing the illustrations and Sally Fuller Reid for preparing the text.

REFERENCES CITED

Acharyya, S. K., Shah, S. C., Ghosh, S. C., and Ghosh, R. N., 1979, Gondwana of the Himalaya and its biostratigraphy, *in* Laskar, B., and Raja Rao, C. S., chief eds., 4th International Gondwana Symposium, Calcutta 1977: Delhi, Hindustan Publishing Corp., v. 2, p. 420–433.

Adloff, M.-C., Doubinger, J., and Jaeger, J.-J., 1984, Une association palynologique d'affinités laurasiatiques dans le Trias supérieur du bloc de Lhassa (Tibet): Société Géologique de France Mémoire, n.s., no. 147, p. 7–8.

Agocs, W. B., Meyerhoff, A. A., and Kis, K., 1992, Reykjanes Ridge: quantitative determinations from magnetic anomalies, *in* Chatterjee, S., and Hotton, N., III, eds., New Concepts in Global Tectonics: Lubbock, Texas Tech University Press, p. 221–238.

Aguirre-Urreta, M. B., Buatois, L. A., Chernoglasov, G. C. B., and Medina, F. A., 1990, First Polychelidae (Crustacea, Palinura) from the Jurassic of Antarctica: Antarctic Science, v. 2, p. 157–162.

Ahlfeld, F., and Branisa, L., 1960, Geología de Bolivia: La Paz, Instituto Boliviano del Petroleo, Editorial Don Bosco, 245 p.

Anan-Yorke, R., 1974, Devonian chitinozoa and acritarchs from exploratory oil wells on the shelf and coastal regions of Ghana, West Africa: Ghana Geological Survey Bulletin 37, 217 p.

Anderson, J. M., and Anderson, H. M., 1985, Palaeoflora of southern Africa: Prodromus of South African megafloras, Devonian to Lower Cretaceous: Pretoria, Botanical Research Institute, and Rotterdam, A. A. Balkema, 423 p.

Anderson, J. M., and Cruickshank, A. R., 1978, The biostratigraphy of the Permian and Triassic Part 5. A review of the classification and distribution of Permo-Triassic tetrapods: Palaeontologia Africana, v. 21, p. 15–44.

Anderson, M. M., Boucot, A. J., and Johnson, J. G., 1966, Devonian terebratulid brachiopods from the Accraian series of Ghana: Journal of Paleontology, v. 40, p. 1365–1367.

Andrianov, V. N., 1966, Verkhnepaleozoyskiye otlozheniya Zapadnogo Verkhoyanya, *in* Soveshchaniya po stratigrafii Severo-Vostoko SSSR, Magadan, Trudy: Moscow, Izdatel'stvo "Nauka," p. 130–160.

Archangelsky, S., 1960, Lycopsida y Sphenopsida del Paleozóico superior de Chubut y Santa Cruz: Acta Geologica Lilloana, v. 3, p. 21–36.

Archangelsky, S., and Sota, E. R. de la, 1960, Contribución al conocimiento de las filices pérmicas de Patagonia extrandina: Acta Geologica Lilloana, v. 3, p. 85–126.

Archbold, N. W., 1981, Permian brachiopods from western Irian Jaya, Indonesia: Bandung, Geological Research and Development Centre, Paleontology Series 2, p. 1–25.

Archbold, N. W., 1982, Correlation of the Early Permian faunas of Gondwana: Implications for the Gondwanan Carboniferous-Permian boundary: Journal of the Geological Society of Australia, v. 29, p. 267–276.

Archbold, N. W., 1983, Permian marine invertebrate provinces of the Gondwanan Realm: Alcheringa, v. 7, p. 59–70.

Archbold, N. W., 1987, South-western Pacific Permian and Triassic marine faunas: Their distribution and implications for terrane identification, *in* Leitch, E. C., and Schneider, E., eds., Terrane Accretion and Orogenic Belts: American Geophysical Union and Geological Society of America, Geodynamics Series, v. 19, p. 119–127.

Archbold, N. W., 1991a, *Trigonotreta* (Spiriferida, Brachiopoda) from the Early Permian of Victoria: Alcheringa, v. 15, p. 321–326.

Archbold, N. W., 1991b, Early Permian Brachiopoda from Irian Jaya: BMR Journal of Australian Geology and Geophysics, v. 12, p. 287–296.

Archbold, N. W., Pigram, C. J., Ratnam, N., and Hakin, S., 1982, Indonesian Permian brachiopod fauna and Gondwana–south-east-Asia relationships: Nature, v. 296, p. 556–558.

Archer, M., Flannery, T. F., Ritchie, A., and Molnar, R. E., 1985, First Mesozoic mammal from Australia—An Early Cretaceous monotreme: Nature, v. 318, p. 363–366.

Arkell, W. J., 1956, Jurassic Geology of the World: New York, Hafner Publishing Co., 806 p.

Ash, S. R., 1979, *Skilliostrobus* gen. nov., a new lycopsid cone from the Early Triassic of Australia: Alcheringa, v. 3, p. 73–89.

Assefa, G., 1988, Potential hydrocarbon-generating rock units within the Phanerozoic sequence of the Ogaden basin, Ethiopia: A preliminary assessment using the Lopatin model: Journal of Petroleum Geology, v. 11, p. 461–472.

Astibia, H., and 17 others, 1990, The fossil vertebrates from Lano (Basque country, Spain): New evidence on the composition and affinities of the Late Cretaceous continental faunas of Europe: Terra Nova, p. 460–466.

Auden, J. B., 1981, India's former crustal neighbours: Indian National Academy of Sciences Proceedings, pt. A, v. 47, p. 588–630.

Audley-Charles, M. G., 1968, The geology of Portuguese Timor: Geological Society of London Memoir 4, 74 p.

Axelrod, D. I., 1944, The Pliocene sequence in central California, *in* Chaney, R. W., ed., Pliocene Floras of California and Oregon: Carnegie Institution of Washington Contributions to Paleontology, Publication 553, p. 207–224.

Ayme, J.-M., 1965, The Senegal salt basin, *in* Salt Basins around Africa: London, Institute of Petroleum, p. 83–90.

Azhar, H. H., 1977, Geology of the Bukit Jaya area, Pahang [unpub. B.Sc. thesis]: University of Malaya Department of Geology, 63 p.

Bahlburg, H., Breitkreuz, C., and Zeil, W, 1987, Paleozoic basin development in northern Chile (21°-27°): Geologische Rundschau, bd. 76, p. 633–646.

Balmé, B. E., 1969, The Triassic system in Western Australia: The APEA Journal, 9, pt. II, p. 67–78.

Barrett, S. F., and Isaacson, P. E., 1988, Devonian paleogeography of South America, *in* McMillan, N. J., Embry, A. F., and Glass, D. J., eds., Devonian of the World, v. I., Regional Syntheses: Canadian Society of Petroleum Geologists Memoir 14, p. 655–667.

Barry, T. H., 1975, Affinities and systematic position of the South African lower Middle Permian dicynodonts (Therapsida: Dicynodontidae), *in* Campbell, K. S. W., ed., Gondwana Geology, Australia 1973: Canberra, Australian National University Press, p. 475–479.

Bassoullet, J.-P., Colchen, M., Gilbert, É., Marcoux, J., Mascle, G., Sutre, É., and Van Haver, Th., 1984, L'orogène himalayen au Crétacé: Société Géologique de France Mémoire, n.s., no. 147, p. 9–20.

Battail, B., 1981, Faunes de vertébrés et individualité paléontologique du Gondwana au Trias inférieur: Société Géologique de France Bulletin, 7e sér., t. 23, p. 571–578.

Battail, B., Beltan, L., and Dutuit, J.-M., 1987, Africa and Madagascar during Permo-Triassic time: The evidence of the vertebrate faunas, *in* McKenzie, G. D., ed., Gondwana Six: Stratigraphy, sedimentology, and paleontology: American Geophysical Union Geophysical Monograph 41, p. 147–155.

Beltan, L., and Tintori, A., 1981, The genus *Saurichthys* (Pisces, Actinopterygii) during the Gondwana period, *in* Cresswell, M. M., and Vella, P., eds., Gondwana Five: Rotterdam, A. A. Balkema, p. 53–59.

Bertrand-Sarfati, J., Moussine-Pouchkine, A., Affaton, P., Trompette, R., and Bellion, Y., 1991, Cover sequences of the West African craton, *in* Dallmeyer, R. D., and Lecorche, J. P., eds., The West African orogens and circum-Atlantic correlatives: Berlin, Springer-Verlag, p. 65–82.

Besairie, H., 1972, Géologie de Madagascar. I, Les terrains sedimentaires: Tananarive, Annales Géologique de Madagascar, fasc. 35, 463 p.

Betekhtina, O. A., and Bogush, O. I., 1984, Glava VII. Perm', *in* Yanshin, A. L., Chief ed., Phanerozoy Sibiri, t. 1, Vend, Paleozoy: Novosibirsk, Izdatelstvo "Nauka", Sibirskoye Otdeleniye, p. 157–172.

Beuf, S., Biju-Duval, B., Charpal, O. de, Rognon, P., Gariel, O., and Bennacef, A., 1971, Les grès du Paléozoïque inférieur au Sahara: Paris, Publications de l'Institut Français de Pétrole, Collection "Science et Technique du Pétrole," no. 18, Éd. Technip, 464 p.

Bhatia, S. B., 1984 [1985], Ostracode faunas of the Indian subcontinent—their paleozoogeographic and paleoecologic implications: Journal of the Palaeontological Society of India, v. 20, p. 1–8.

Bhatia, S. B., and Rana, R. S., 1984, Palaeogeographic implications of the Charophyta and Ostracoda of the [Deccan] Inter-trappean beds of Peninsular India: Société Géologique de France Mémoire, n.s., no. p. 147, p. 29–35.

Bhattacharya, N., 1981, Depositional patterns in limestones of the Kota Formation (Upper Gondwana), Andhra Pradesh, India, *in* Cresswell, M. M., and Vella, P., eds., Gondwana Five: Rotterdam, A. A. Balkema, p. 135–139.

Bose, M. N., Taylor, E. L., and Taylor, T. N., 1989, Gondwana floras of India and Antarctica—a survey and appraisal, *in* Taylor, T. N., and Taylor, E. L., eds., Antarctic Paleobiology, Its Role in the Reconstruction of Gondwana: New York, Springer-Verlag, p. 119—148.

Boucot, A. J., 1975, Evolution and Extinction Rate Controls: Amsterdam, Elsevier Publishing Co., 427 p.

Boucot, A. J., 1985, Late Silurian–Early Devonian biogeography, provincialism, evolution and extinction: Philosophical Transactions of the Royal Society, London, ser. B, v. 309, p. 323–339.

Boucot, A. J., 1988, Devonian biogeography: An update, *in* McMillan, N. J., Embry, A. F., and Glass, D. J., ed., Proceedings, 2nd International Symposium on the Devonian System: Canadian Society of Petroleum Geologists Memoir 14, v. III, p. 211–227.

Boucot, A. J., 1990, Silurian biogeography, *in* McKerrow, W. S., and Scotese, C. R., eds., Palaeozoic palaeogeography and biogeography: Geological Society of London Memoir no. 12, p. 191–196.

Boucot, A. J., and Gray, J., 1979, Epilogue: A Paleozoic Pangaea?, *in* Gray, J., and Boucot, A. J., eds., Historical Biogeography, Plate Tectonics, and the Changing Environment: Corvallis, Oregon State University Press, p. 465–482.

Boucot, A. J., and Gray, J., 1983, A Paleozoic Pangaea: Science, v. 222, p. 571–581.

Boucot, A. J., Johnson, J. G., and Talent, J. A., 1969, Early Devonian brachiopod zoogeography: Geological Society of America Special Paper 119, 107 p.

Boucot, A. J., Isaacson, P. E., and Laubacher, G., 1980, An Early Devonian, Eastern Americas Realm faunule from the coast of southern Peru: Journal of Paleontology, v. 54, p. 359–365.

Boucot, A. J., Brunton, C. H. C., and Theron, J. N., 1983, Implications for the age of South African rocks in which *Tropidoleptus* (Brachiopoda) has been found: Geological Magazine, v. 120, p. 51–58.

Boucot, A. J., Rohr, D. M., and Blodgett, R. B., 1988, A marine invertebrate faunule from the Tawil Sandstone (basal Devonian) of Saudi Arabia and its biogeographic-paleogeographic consequences: New Mexico Department of Mines Memoir 44, p. 361–372.

Boutakoff, N., 1940, The Dwyka glaciation and the Ecca epiglacial beds of the Congo basin, *in* Symposium on Palaeozoic and Pre-Cambrian Climates; 17th International Geological Congress: Moscow, 1937, Report 6, p. 223–245.

Boutakoff, N., 1948, Les formations glacieres et post-glacieres fossiliferes, d'age Permo-Carbonifere (Karroo inferieur) de la région de Walikale (Kivu, Congo Belge): Université Louvain, Institut Géologique, Mémoire 9, 124 p.

Bowen, R. L., 1964, Isolation of late Paleozoic glaciations of eastern Australia, *in* Mehta, D. R. S., and Ahmad, F., eds., Gondwanas, 22nd International Geological Congress; New Delhi, 1964, Report 9, p. 73–89.

Braakman, J. H., Levell, B. K., Martin, J. H., Potter, T. L., and Vliet, A. van, 1982, Late Palaeozoic Gondwana glaciation in Oman: Nature, v. 299, p. 48–50.

Brasier, M. D., and Singh, P., 1987, Microfossils and Precambrian-Cambrian boundary stratigraphy at Maldeota, Lesser Himalaya: Geological Magazine, v. 124, p. 323–345.

Brookfield, M. E., 1987, Stratigraphic correlation of Blaini Formation (late Proterozoic, Lesser Himalaya, India) with other late Proterozoic tillite sequences: Geologische Rundschau, bd. 76, p. 477–484.

Brooks, C. E. P., 1926, Climate through the Ages: New Haven, Connecticut, Yale University Press, 439 p.

Brooks, C. E. P., 1949, Climate through the ages (rev. ed.): New York, McGraw-Hill, 395 p.

Brouwers, E. M., Clemens, W. A., Spicer, R. A., Ager, T. A., Carter, L. D., and Sliter, W. V., 1987, Dinosaurs on the North Slope, Alaska: High latitude, latest Cretaceous environments: Science, v. 237, p. 1608–1610.

Brown, D. A., Campbell, K. S. W., and Crook, K. A. W., 1968, The geological evolution of Australia and New Zealand: London, Pergamon Press, 409 p.

Bucher, W. H., 1964, The third confrontation, *in* Nairn, A. E. M., ed., Problems in Palaeoclimatology: London, Wiley-Interscience, p. 3–9.

Buffetaut, E., 1989, The contribution of vertebrate palaeontology to the geodynamic history of south east Asia, *in* Sengör, A. M. C., ed., Tectonic Evolution of the Tethyan Region (NATO ASI Series): Dordrecht, Kluwer Academic Publishers, p. 645–653.

Buffetaut, E., and Ingavat, R., 1986, The succession of vertebrate faunas in the continental Mesozoic of Thailand: Geological Society of Malaysia Bulletin, no. 19, p. 167–172.

Burckle, L. H., and Pokras, E. M., 1991, Implications of a Pliocene stand of *Nothofagus* (southern beech) within 500 kilometers of the South Pole: Antarctic Science, v. 3, p. 389–403.

Busson, G., 1970, Le Mésozoïque saharien, 2e partie: Essai de synthèse des données des sondages algéro-tunisiens, T. 1: Paris, Centre de Recherches sur les Zones Arides, Série, Géologie no. 11, Éditions du Centre National de la Recherche Scientifique, 340 p.

Cain, S. A., 1944, Foundations of Plant Geography: New York, Harper & Brothers, 556 p.

Camp, C. L., and Banks, M. R., 1978, A proterosuchian reptile from the Early Triassic of Tasmania: Alcheringa, v. 2, p. 143–158.

Cappetta, H., 1987, Chondrichthyes II. Mesozoic and Cenozoic Elasmobranchii, *in* Schultze, H. P., ed., Handbook of Paleoichthyology, v. 3B: Stuttgart, G. Fischer, 193 p.

Cariou, E., 1973, Ammonites of the Callovian and Oxfordian, *in* Hallam, A., ed., Atlas of Palaeobiogeography: Amsterdam, Elsevier Publishing Co., p. 287–295.

Castaños, A., and Rodrigo-G., L.A., 1978, Sinópsis estratigráfico de Bolivia, I parte: Paleozóico: La Paz, Academia Nacional de Ciencias de Bolivia, 146 p.

Cecioni, G., 1955, Noticias preliminares sobre el hallazgo del paleozóico superior en el archipiélago patagónico: Universidad de Chile, Facultad de Ciencias, Anales, v. 12, p. 257–258.

Chaloner, W. G., and Lacey, W. S., 1973, The distribution of late Palaeozoic floras, *in* Hughes, N. F., ed., Organisms and Continents through Time: Palaeontological Association, Special Papers in Palaeontology, no. 12, p. 271–289.

Chaloner, W. G., and Meyen, S. V., 1973, Carboniferous and Permian floras of the northern continents, *in* Hallam, A., ed., Atlas of Palaeobiogeography: Amsterdam, Elsevier Publishing Co., p. 169–186.

Chamot, G. A., 1965, Permian section at Apillapampa, Bolivia and its fossil content: Journal of Paleontology, v. 39, p. 1112–1124.

Chandra, S., and Surange, K. R., 1979, Revision of the Indian species of *Glossopteris*: Lucknow, Birbal Sahni Institute of Palaeobotany Monograph 2, 291 p.

Charig, A. J., 1971, Faunal provinces on land: Evidence based on the distribution of fossil tetrapods, with especial reference to the reptiles of the Permian and Mesozoic, *in* Middlemiss, F. A., and Rawson, P. F., eds., Faunal Provinces in Space and Time: Liverpool, Seel House Press, p. 111–128.

Charig, A. J., 1973, Jurassic and Cretaceous dinosaurs, *in* Hallam, A., ed., Atlas of Palaeobiogeography: Amsterdam, Elsevier Publishing Co., p. 339–352.

Chatterjee, S., 1984, A new ornithischian dinosaur from the Triassic of North America: Naturwissenschaften, bd. 71, p. 630–631.

Chatterjee, S., 1986a, *Malerisaurus langstoni*, a new diapsid reptile from the Triassic of Texas: Journal of Vertebrate Paleontology, v. 6, p. 297–312.

Chatterjee, S., 1986b, The Late Triassic vertebrates: Their stratigraphic and paleobiogeographic significance, *in* Padian, K., ed., The Beginning of the Age of Dinosaurs: Cambridge, Cambridge University Press, p. 139–150.

Chatterjee, S., and Hotton, N., III, 1986, The paleoposition of India: Journal of Southeast Asian Earth Sciences, v. 1, no. 3, p. 145–189.

Chen Bingwei, 1982, The discovery of Middle and Upper Carboniferous tilloids in Baxoi, Xizang (Tibet) and their significance: Geological Review, v. 28, no. 2, p. 148–152.

Chen Tingen, 1983, The discovery of *Georgina* Wade from southern Xizang (Tibet), and its significance: Nanjing Institute of Geology and Palaeobiology Bulletin, no. 6, p. 129–130 (English resume).

Cheng Zhengwu, 1981, Permo-Triassic continental deposits and vertebrate faunas of China, *in* Cresswell, M. M., and Vella, P., eds., Gondwana Five: Rotterdam, A. A. Balkema, p. 65–70.

Chiplonkar, G. W., and Phansalkar, V. G., 1980, A study of mortoniceratids from the Utatur rocks of south India: Recent Researches in Geology, Hindustan Publishing Corp. (Delhi), v. 6, p. 54–464.

Chonglakmani, C., 1983, The marine Mesozoic stratigraphy of Thailand, *in* Nutalaya, P.,ed.-in-chief, Proceedings, Workshop on Stratigraphic Correlation in Thailand and Malaysia: Geological Society of Thailand and Geological Society of Malaysia, p. 105–126.

Clarke, J. M., 1913, Fosseis devonianos do Paraná: Serviço Geológico e Mineralógico do Brasil, Monografia 1, 353 p.

Clarke, M. J., 1990, Late Palaeozoic (Tamarian; Late Carboniferous–Early Permian) cold-water brachiopods from Tasmania: Alcheringa, v. 14, p. 53–76.

Cluver, M. A., and Hotton, N., III, 1979, The dicynodont genus *Diictodon* (Reptilia, Therapsida) and its significance, *in* Laskar, B., and Raja Rao, C.S., eds., 4th International Gondwana Symposium, Calcutta 1977: Delhi, Hindustan Publishing Corporation, v. 1, p. 176–183.

Colbert, E. H., 1938, Fossil mammals from Burma in the American Museum of Natural History: American Museum of Natural History Bulletin, v. 76, p. 255–436.

Colbert, E. H., 1969, Evolution of the Vertebrates. A History of the Backboned Animals through time (second edition): New York, John Wiley & Sons, 535 p.

Colbert, E. H., 1979, Gondwana vertebrates, *in* Laskar, B., and Raja Rao, C. S., chief eds., 4th International Gondwana Symposium, Calcutta, 1977: Delhi, Hindustan Publishing Corp., p. 135–143.

Colbert, E. H., 1981, The distribution of tetrapods and the break-up of Gondwana, *in* Cresswell, M. M., and Vella, P., eds., Gondwana Five: Rotterdam, A. A. Balkema, p. 277–282.

Colbert, E. H., 1983, Triassic vertebrates in the Transantarctic Mountains, *in* Turner, M. D., and Splettstoesser, J. E., eds., Geology of the Central Transantarctic Mountains: American Geophysical Union Antarctic Research Series, v. 36, paper 2, p. 11–35.

Cooper, M. R., 1980, Palaeobiogeography and the reassembly of Gondwanaland: South African Journal of Science, v. 76, p. 35–37.

Cooper, M. R., and Kensley, B., 1984, Endemic South American Permian bivalve molluscs from the Ecca of South Africa: Journal of Paleontology, v. 58, p. 1360–1363.

Coquel, R., 1985 [1987], Microfloras, *in* C. Martínez-Díaz, ed., North Africa, The Carboniferous of the World, IUGS Publication 20, II, Australia, Indian Subcontinent, South Africa, South America, and North Africa: Madrid, Instituto Geológico y Minero de España, p. 381–386.

Cortés, J. M., Caminos, R., and Leanza, H. A., 1984, La cobertura sedimentaria eopaleozóica, *in* IX Congreso Geológico Argentina, San Carlos de Bariloche, Relatorio: Buenos Aires, Congresos Geológicos Argentinos, p. 65-84.

Cosgriff, J. W., 1965, A new genus of Temnospondyli from the Triassic of Western Australia: Journal of the Royal Society of Western Australia, v. 48, pt. 3, p. 65–95.

Cosgriff, J. W., 1969, *Blinasaurus*, a brachyopid genus from Western Australia and New South Wales: Journal of the Royal Society of Western Australia, v. 52, pt. 3, p. 65–88.

Cosgriff, J. W., 1972, *Parotosaurus wadei*, a new capitosaurid from New South Wales: Journal of Paleontology, v. 46, p. 545–555.

Cosgriff, J. W., 1974, Lower Triassic Temnospondyli of Tasmania: Geological Society of America Special Paper 149, 134 p.

Cosgriff, J. W., and DeFauw, S. L., 1987, A capitosaurid labyrinthodont from the early Scythian of Tasmania: Alcheringa, v. 11, p. 22–41.

Cosgriff, J. W., and Hammer, W. R., 1983, The labyrinthodont amphibians of the earliest Triassic from Antarctica, Tasmania and South Africa, *in* Oliver, R. L., James, P. R., and Jago, J. B., eds., Antarctic Earth Science: Cambridge, Cambridge University Press, p. 611–618.

Cosgriff, J. W., Hammer, W. R., and Ryan, W. J., 1982, The Pangaean reptile, *Lystrosaurus maccaigi*, in the Lower Triassic of Antarctica: Journal of Paleontology, v. 56, p. 371–385.

Cousminer, H. L., 1965, Permian spores from Apillapampa, Bolivia: Journal of Paleontology, v. 39, p. 1097–1111.

Cox, C. B., 1973, Triassic tetrapods, *in* Hallam, A., ed., Atlas of Palaeobiogeography: Amsterdam, Elsevier Publishing Co., p. 213–223.

Cox, L. R., 1936, Karroo Lamellibranchia from Tanganyika Territory and Madagascar: Geological Society of London Quarterly Journal, v. 92, pt. 1, p. 32–57.

Crame, J. A., 1981, The occurrence of *Anopaea* (Bivalvia: Inoceramidae) in the Antarctic Peninsula: Journal of Molluscan Studies, v. 47, p. 206–219.

Crame, J. A., 1982, Late Mesozoic bivalve biostratigraphy of the Antarctic Peninsula region: Journal of the Geological Society of London, v. 139, p. 771–778.

Crame, J. A., 1986, Late Mesozoic bipolar bivalve faunas: Geological Magazine, v. 123, p. 611–618.

Crame, J. A., 1987, Late Mesozoic bivalve biogeography of Antarctica, *in* McKenzie, G. D., ed., Gondwana Six: American Geophysical Union Geophysical Monograph 41, p. 93–102.

Crawford, A. R., 1974, The Indus suture line, the Himalaya, Tibet and Gondwanaland: Geological Magazine, v. 111, p. 369–383.

Culver, S. J., Pojeta, J., Jr., and Repetski, J. E., 1988, First record of Early Cambrian shelly microfossils from West Africa: Geology, v. 16, no. 7, p. 596–599.

Culver, S. J., Williams, H. R., and Venkatakrishnan, R., 1991, The Rokelide orogen, *in* Dallmeyer, R. D., and Lécorché, J. P., eds., The West African orogens and circum-Atlantic correlatives: Berlin, Springer-Verlag, p. 123–150.

Cyrtoký, P., 1973, Permian flora from the Ga'ara region (western Iraq): Neues Jahrbuch für Geologie und Paläontologie, Monatshefte 1973, no. 7, p. 383–388.

Dagis, A. S., 1974, Triassic brachiopods (morphology, classification, phylogeny, stratigraphical significance and biogeography): Novosibirsk, Izdatel'stvo "Nauka," Sibirskoye Otdel, 385 p.

Dagis, A. S., and Ustritskiy, V. I., 1973, The main relationships between changes in marine fauna at the close of the Permian and the beginning of the Triassic, *in* Logan A., and Hills, L. V., eds., The Permian and Triassic Systems and Their Mutual Boundary: Canadian Society of Petroleum Geologists Memoir 2, p. 647–654.

Dallmus, K. F., 1958, Mechanics of basin evolution and its relation to the habitat of oil in the basin, *in* Weeks, L. G., ed., Habitat of Oil: Tulsa, Oklahoma, American Association of Petroleum Geologists, p. 883–931.

Daly, M. C., Lawrence, S. R., Diemu-Tshiband, K., and Matouana, B., 1992, Tectonic evolution of the Cuvette Centrale, Zaire: Geological Society of London Journal, v. 149, pt. 4, p. 539–546.

Dehm, R., and Oettingen-Spielberg, T. zu, 1958, Paläontologische und geologische Untersuchungen im Tertiär von Pakistan. 2, Die mitteleocänen Säugetiere von Ganda Kas bei Basal in Nordwest-Pakistan: Bayerische Akademie der Wissenschaften, matematisch-naturwissenschaftliche Klasse, Abhandlungen, Heft 91, 54 p.

Oliveira, A. I. de, 1956, Brazil, *in* Jenks, W. F., ed., Geological Society of America Memoir 65, p. 1–62.

de Lapparent, A. F., 1962, Footprints of dinosaurs in the Lower Cretaceous of Vestspitsbergen-Svalbard: Norsk Polarinstitutt Arbok, 1960, p. 14–21.

de Lapparent, A. F., Termier, G., and Termier, H., 1971, Découverté de la fauna d'Umaria (Permien inferieur de l'Inde) en Afghanistan et consequences paléogéographiques: Comptes Rendus Academie de Science, sér. D, t. 272, p. 381–384.

Delevoryas, T., 1969, Glossopterid leaves from the Middle Jurassic of Oaxaca, Mexico: Science, v. 165, p. 895–896.

Delevoryas, T., and Person, C. P., 1975, *Mexiglossa varia* gen. et sp. nov., a new genus of glossopterid leaves from the Jurassic of Oaxaca, Mexico: Palaeontographica, abt. B, bd. 154, p. 114–120.

Derby, O. A., 1894, The Amazonian Upper Carboniferous fauna: Journal of Geology, v. 2, p. 480–501.

De Wever, P., Bourdillon-de Grissac, C., and Bechennec, F., 1988, Permian age from radiolarites of the Hawasina nappes, Oman Mountains: Geology, v. 16, p. 912–914.

Dewey, J. F., 1988, Extensional collapse of orogens: Tectonics, v. 7, p. 1123–1139.

Dickins, J. M., 1961, *Eurydesma* and *Peruvispira* from the Dwyka beds of South Africa: Palaeontology, v. 4, p. 138–148.

Dickins, J. M., 1973, The geological sequence and the Permian-Triassic boundary in Australia and eastern New Guinea, *in* Logan, A., and Hills, L.V., eds., The Permian and Triassic Systems and Their Mutual Boundary: Canadian Society of Petroleum Geologists Memoir 2, p. 425–432.

Dickins, J. M., 1978, Climate of the Permian in Australia: The invertebrate faunas: Palaeogeography, Palaeoclimatology, Palaeoecology, v. 23, p. 33–46.

Dickins, J. M., 1984, Evolution and climate in the Upper Palaeozoic, *in* Brenchley, P., ed., Fossils and Climate: Chichester, UnitedKingdom, John Wiley & Sons, p. 317–327.

Dickins, J. M., 1985a, Late Palaeozoic glaciation: BMR, Journal of Australian Geology and Geophysics, v. 9, no. 2, p. 163–169.

Dickins, J. M., 1985b, Palaeobiofacies and palaeobiogeography of Gondwanaland from Permian to Triassic, *in* Nakazawa, K., and Dickins, J. M., eds., The Tethys: Tokyo, Tokai University Press, p. 83–92.

Dickins, J. M., 1985c, Climate of the Triassic, *in* Hornibrook Symposium, 1985, extended abstracts, New Zealand Geological Survey Record, v. 9, p. 34–36.

Dickins, J. M., 1985d, Late Paleozoic climate with special reference to invertebrate faunas: Compte Rendu Neuvième Congrès International de Stratigraphie et de Géologie du Carbonifère, Washington, Urbana, v. 5, p. 394–402.

Dickins, J. M., and McTavish, R. A., 1963, Lower Triassic marine fossils from the Beagle Ridge (B.M.R. 10) bore, Perth basin, Western Australia: Journal of the Geological Society of Australia, v. 10, pt. 1, p. 123–140.

Dickins, J. M., and Shah, S. C., 1979, Correlation of the Permian marine sequences of India and Western Australia, *in* Laskar, B., and Raja Rao, C.S., eds., 4th International Gondwana Symposium, Calcutta, 1977: Delhi, Hindustan Publishing Corporation, v. 2, p. 387–408.

Dickins, J. M., and Shah, S. C., 1981, Permian paleogeography of Peninsular and Himalayan India and the relationship with the Tethyan region, *in* Cresswell, M. M., and Vella, P., eds., Gondwana Five: Rotterdam, A. A. Balkema, p. 79–83.

Dickins, J. M., Archbold, N. W., Thomas, G. A., Campbell, H. J., 1989, Mid-Permian correlation: Compte Rendu Onzième Congrès International de Stratigraphie et de Géologie du Carbonifère, Beijing, v. 2, p. 185–198.

Dickins, J. M., Shah, S. C., Archbold, N. W., Jin Yugan, Liang Dinyi, and Liu Benpei, 1993, Some climatic and tectonic implications of the Permian marine faunas of Peninsular India, Himalayas and Tibet, *in* Findlay, R.H., Unrug, R., Banks, M.R., and Veevers, J. J., eds., Gondwana Eight—Assembly, Evolution and Dispersal: Rotterdam, A. A. Balkema, p. 333–343.

Dietz, R. S., and Holden, J. C., 1970, Reconstruction of Pangaea: Breakup and dispersion of continents, Permian to present: Journal of Geophysical Research, v. 75, p. 4939–4956.

Dingle, R. V., and Klinger, H. C., 1971, Significance of Upper Jurassic sediments in the Knysna Outlier (Cape Province) for timing of the breakup of Gondwanaland: Nature, Physical Sciences, v. 232, p. 37–38.

Dingle, R. V., Siesser, W. G., and Newton, A. R., 1983, Mesozoic and Tertiary geology of southern Africa: Rotterdam, A. A. Balkema, 375 p.

Dolianiti, E., 1972, Relações entre as floras paleozóicas do Brasil: Academia Brasileira Ciencias Anais, v. 44 (Suplemento), p. 113–118.

Douglass, R. C., and Nestell, M. K., 1976, Late Paleozoic foraminifera from southern Chile: U.S. Geological Survey Professional Paper 858, 49 p.

Dow, D. B., Beyth, M., and Hailu, T., 1971, Palaeozoic glacial rocks recently discovered in northern Ethiopia: Geological Magazine, v. 108, pt. 1, p. 53–60.

Drewry, G. E., Ramsay, A. T. S., and Smith, A. G., 1974, Climatically controlled sediments, the geomagnetic field, and trade wind belts in Phanerozoic time: Journal of Geology, v. 82, p. 531–554.

Drysdall, A. R., Johnson, R. L., Moore, T. A., and Thieme, J. G., 1972, Outline of the geology of Zambia: Geologie en Mijnbouw, v. 51, p. 265–276.

Dunbar, C. O., and Newell, N. D., 1946, Marine Early Permian of the central

Andes and its fusuline faunas: American Journal of Science, v. 244, pt. I, p. 377–402; pt. II, p. 457–491.

Toit, A. L. du, 1954, The geology of South Africa, (third edition, rev.): New York, Hafner Publishing Co., 611 p.

Dutro, J. T., Jr., and Saldukas, R. B., 1973, Permian paleogeography of the Arctic: Journal of Research of the U.S. Geological Survey, v. 1, p. 501–507.

Dutt, A. B., and Shah, S. C., 1969, Marine fossils from the Talchir Series of the Daltonganj coalfield, Bihar: Geological Survey of India Records, v. 97, pt. 2, p. 102.

Edwards, C. W., 1982, New paleontologic evidence of Triassic sedimentation in West Antarctica, *in* Craddock, C., ed., Antarctic Geoscience: International Union of Geological Sciences, ser. B, no. 4, University of Wisconsin Press, p. 325–330.

El-Khayal, A. A., and Romano, M., 1988, A revision of the upper part of the Saq Formation and Hanadir shale (Lower Ordovician) of Saudi Arabia: Geological Magazine, v. 129, p. 161–174.

El-Khayal, A. A., Chaloner, W. G., and Hill, C. R., 1980, Palaeozoic plants from Saudi Arabia: Nature, v. 285, p. 33–34.

Engel, B. A., and Morris, L. N., 1985 [1987], Trilobites of eastern Australia, *in* Martínez-Díaz, C., ed., The Carboniferous of the World, International Union of Geological Sciences Publication 20. II—Australia, Indian Subcontinent, South Africa, South America, and North Africa: Madrid, Instituto de Geología y Minero de España, p. 108–112.

Epshteyn, O. G., 1981a, Middle Carboniferous ice-marine deposits of northeastern U.S.S.R., *in* Hambrey, M. J., and Harland, W. B., eds., Earth's Pre-Pleistocene Glacial Record: Cambridge, Cambridge University Press, p. 268–269.

Epshteyn, O. G., 1981b, Late Permian ice-marine deposits of the Atkan Formation in the Kolyma River headwaters region, *in* Hambrey, M. J., and Harland, W. B., eds., Earth's Pre-Pleistocene Glacial Record, Cambridge, Cambridge University Press, p. 270–273.

Evans, K. R., and Rowell, A. J., 1990, Small shelly fossils from Antarctica: An Early Cambrian faunal connection with Australia: Journal of Paleontology, v. 64, p. 692–700.

Fairbridge, R. W., 1965, The Indian Ocean and the status of Gondwanaland: Progress in Oceanography, v. 3, p. 83–136.

Fang, T. C., Xia, T. L., and Liu, H. L., 1979, Paleontological stratigraphy: Beijing, Geological Press, 308 p.

Fang Zongjie, 1993, Biogeographic restraints on the rift-drift-accretion history of the Sibumasa block: Journal of Southeast Asian Earth Sciences, v. 9, no. 4, p. 375–385.

Fischer, A. G., 1960, Latitudinal variations in organic diversity: Evolution, v. 14, p. 50–73.

Flint, R. F., 1947, Glacial geology and the Pleistocene Epoch: New York, John Wiley & Sons, 589 p.

Florin, R., 1963, The distribution of conifer and taxad genera in time and space: Uppsala, Acta Horti Bergiani, Bd. 20, p. 120–280.

Fontaine, H., 1986, Discovery of Lower Permian corals in Sumatra: Geological Society of Malaysia Bulletin, no. 19, p. 183–191.

Fontaine, H., and Vachard, D., 1984, New paleontological data on the upper Paleozoic of Sumatra: Société Géologique de France Mémoire, n.s., no. 147, p. 49–53.

Foster, C. B., Balmé, K. K., and Watson, S. T., 1994, First record of Tethyan palynomorphs from the Late Triassic of East Antarctica: AGSO Journal of Australian Geology and Geophysics, v. 15, p. 239–246.

Francis, J. E., and McMillan, N. J., 1987, Fossil forests in far north: Geos, v. 16, p. 6–9.

Franks, S., and Nairn, A. E. M., 1973, The equatorial marginal basins of West Africa, *in* Nairn, A. E. M., and Stehli, F. G., eds., The Ocean Basins and Margins, v. 1. The South Atlantic: New York, Plenum Press, p. 301–350.

Funder, S., Abrahamsen, N., Bennike, O., and Feyling-Hanssen, R.W., 1985, Forested Arctic: Evidence from north Greenland: Geology, v. 13, p. 542–546.

Furon, R., 1968, Géologie de l'Afrique, (third edition): Paris, Payot, 374 p.

Galton, P. M., 1977, The ornithopod dinosaur *Dryosaurus* and a Laurasia-Gondwanaland connection in the Upper Jurassic: Nature, v. 268, p. 230–232.

Gansser, A., 1964 [1965], Geology of the Himalayas: London, Interscience Publishers, 289 p.

Gansser, A., 1981, The geodynamic history of the Himalaya, *in* Gupta, H. K., and Delany, F. M., eds., Zagros-Hindu Kush-Himalaya Geodynamic Evolution: American Geophysical Union and Geological Society of America, Geodynamics Series, v. 3, p. 111–121.

Gansser, A., 1983, Geology of the Bhutan Himalaya: Schweizerischen Naturforschenden Gesellschaft, Denkschriften, bd. 96, 181 p.

Gansser, A., 1991, Facts and theories on the Himalayas: Eclogae Geologicae Helvetiae, v. 84, p. 33–59.

Gao Keqin, 1989, Pareiasaurs from the Upper Permian of North China: Canadian Journal of Earth Sciences, v. 26, p. 1234–1240.

Gardiner, B. G., 1962, *Namaichthys schroederi* Gurich and other Palaeozoic fishes from South Africa: Palaeontology, v. 5, p. 9–21.

Gardiner, B. G., 1969, New palaeonicoid fish from Witteberg Series of South Africa: Zoological Journal, Linnean Society (London), v. 48, p. 423–452.

Gayet, M., Rage, J.-C., and Rana, R. S., 1984, Nouvelles ichtyofaune et herpetofaune de Gitti Khedan le plus ancien gisement connu du Deccan (Crétacé/Paléocéne) a microvertébrés. Implications paléogéographiques: Société Géologique de France Mémoire, n.s., no. 147, p. 55–65.

Germs, G. J. B., 1972, Trace fossils from the Nama Group, South-West Africa: Journal of Paleontology, v. 46, p. 864–870.

Germs, G. J. B., 1973, Possible sprigginid worm and a new trace fossil from the Nama Group, South West Africa: Geology, v. 1, p. 69–70.

Gerth, H., 1957, Das Vorkommen von permokarbonischen Fusulinenkalken im west-patagonischen Archipel und seine paläogeographische und palaoklimatologische Bedeutung: Zeitschrift der Deutschen Geologischen Gesellschaft, bd. 100, teil 1, p. 193–198.

Gobbett, D. J., 1967, Palaeozoogeography of the Verbeekinidae (Permian foraminifera), *in* Adams, C. G., and Ager, D. V., eds., Aspects of Tethyan Biogeography; London, United Kingdom, Systematics Association Publication no. 7, p. 77–90.

Gobbett, D. J., 1968, The Permian System in Malaya: Geological Society of Malaysia Bulletin, no. 1, p. 17–22.

Gobbett, D. J., 1973a, Permian Fusulinacea, *in* Hallam, A., ed., Atlas of Palaeobiogeography: Amsterdam, Elsevier Publishing Co., p. 151–158.

Gobbett, D. J., 1973b, Carboniferous and Permian stratigraphic correlation in southeast Asia: Geological Society of Malaysia Bulletin, no. 6, p. 131–142.

Gonzalez, C. R., 1990, Development of the late Paleozoic glaciations of the South American Gondwana in western Argentina: Palaeaogeography, Palaeoclimatology, Palaeoecology, v. 79, p. 275–287.

Gothan, W., and Weyland, H., 1954, Lehrbuch der Palaotektonik: Berlin, Akademie Verlag, 316 p.

Govindan, A., 1975, Jurassic freshwater ostracods from the Kota Limestone of India: Palaeontology, v. 18, p. 207–216.

Grant, R. E., 1976, Permian brachiopods from southern Thailand: Paleontological Society Memoir 9, Journal of Paleontology, v. 50, suppl. to no. 3, 269 p.

Gray, J., 1988, Land plant spores and the Ordovician-Silurian boundary: British Museum (Natural History), Bulletin, Geology, v. 43, p. 351–358.

Grekoff, N., 1957, Ostracodes de bassin du Congo: I, Jurassique supérieur et Crétacé inférieur du nord du bassin: Musee Royal Congo Belge, Annales, Sciences Géologiques, v. 19, 97 p.

Grekoff, N., 1963, Contribution a l'étude des ostracodes du Mésozoiques moyen (Bathonien-Valanginien) du bassin du Majunga, Madagascar: Revue de l'Institut Français du Pétrole, Annales, v. 18, p. 1709–1762.

Grunt, T. A., and Dmitriyev, V. V., 1973, Perm'skiye brakhiopody Pamira: Akademia Nauk SSSR, Paleontologicheskaya Instituta, Trudy, v. 136, 212 p.

Guimaraes, D., 1964, Geología do Brasil: Rio de Janeiro, Ministerio das Minas e Energía, Memorial, 674 p.

Gu Zhiwei, Sha Jingeng, Li Zishun, and Yu Jinshan, 1984, Occurrence of marine Jurassic bivalves in eastern Northeast China and its significance on the non-marine Jurassic-Cretaceous boundary in east and central Asia, *in* Su Zongwei, ed., Contribution to 17th International Geological Congress, 1984, Moscow: Beijing, Science Press, Developments in Geoscience, p. 111–118.

Habicht, J. K. A., 1979, Paleoclimate, Paleomagnetism, and Continental Drift: Tulsa, Oklahoma, American Association of Petroleum Geologists Studies in Geology, no. 9, 33 p.

Hallam, A., 1972, Continental drift and the fossil record: Scientific American, v. 227, p. 55–66.

Hallam, A., 1989, Great Geological Controversies (second edition): Oxford, United Kingdom, Oxford University Press, 244 p.

Haller, J., 1979, Himalayan orogenesis in perspective, *in* Verma, P. K., ed., Metamorphic rock sequences of the eastern Himalaya: Calcutta, K. P. Bagchi and Company, p. i–xxxiv.

Ham, C. K., and Herrera, L. J., Jr., 1963, Role of Subandean fault system in tectonics of eastern Peru and Ecuador, *in* Childs, O. E., and Beebe, B. W., eds., Backbone of the Americas: Tulsa, Oklahoma, American Association of Petroleum Geologists Memoir 2, p. 47–61.

Hammer, W. R., 1987, Paleoecology and phylogeny of the Trematosauridae, *in* McKenzie, G. D., ed., Gondwana Six: Stratigraphy, Sedimentology, and Paleontology: American Geophysical Union Geophysical Monograph 41, p. 73–83.

Hammer, W. R., 1989, Triassic terrestrial vertebrate faunas of Antarctica, *in* Taylor, T. N., and Taylor, E. L., eds., Antarctic Paleobiology, Its Role in the Reconstruction of Gondwana: New York, Springer-Verlag, p. 42–50.

Hammer, W. R., and Hickerson, W.J., 1992, Comments on the fossil vertebrates from the Falla Formation Jurassic, Beardmore Glacier region, Antarctica: Antarctic Journal of the U.S., v 27, no. 5, p. 1.

Hammer, W. R., Collinson, J. W., and Ryan, W. J., III, 1990, A new Triassic vertebrate fauna from Antarctica and its depositional setting: Antarctic Science, v. 2, p. 163–167.

Han Neiren, Ouyang Chengfu, Li Wenhua, and Li Ruliang, 1991, New views on the Carbo-Permian strata in Laochang, Lancang areas of Yunnan: Journal of Stratigraphy (Beijing), v. 15, p. 56–58.

Hardie, L. A., 1990, The roles of rifting and hydrothermal $CaCl_2$ brines in the origin of potash evaporites: An hypothesis: American Journal of Science, v. 290, p. 1–42.

Hardie, L. A., 1991, On the significance of evaporites: Annual Review of Earth and Planetary Sciences: v. 19, p. 131–168.

Harland, W. B., Armstrong, R. L., Cox, A. V., Craig, L. E., Smith, A. G., and Smith, D. G., 1990, A Geologic Time Scale 1989: Cambridge, Cambridge University Press, 262 p.

Harrington, H. J., 1962, Paleogeographic development of South America: American Association of Petroleum Geologists Bulletin, v. 46, p. 1773–1814.

Hart, G. F., 1964, Where was the Lower Karroo sea?: Scientific South Africa, June 1964, p. 289–290.

Haughton, S. H., 1963, The stratigraphic history of Africa south of the Sahara: New York, Hafner Publishing Co., 365 p.

Haughton, S. H., 1969, Geological history of southern Africa: Cape Town, Geological Society of South Africa, 535 p.

He Xinye, and Weng Fa, 1983, New material of Early Permian corals from Ali, northern Xizang (Tibet): Wuhan College of Geology Earth Science Journal, total 19, no. 1, p. 69–78.

Helby, R., Wiggins, V. D., and Wilson, G. J., 1987, The Circum-Pacific occurrence of the Late Triassic dinoflagellate *Sverdrupiella*: Australian Journal of Earth Sciences, v. 34, p. 151–152.

Hill, D., 1958, Sakmarian geography: Geologische Rundschau, bd. 47, p. 590–629.

Hill, D., 1973, Lower Carboniferous corals, *in* Hallam, A., ed., Atlas of Palaeobiogeography: Amsterdam, Elsevier Publishing Co., p. 133–142.

Hiller, N., and Theron, J. N., 1988, Benthic communities in the South African Devonian, *in* McMillan, N. J., Embry, A. F., and Glass, D. J., eds., Devonian of the World, v. III, Paleontology, Paleoecology and Biostratigraphy: Canadian Society of Petroleum Geologists Memoir 14, p. 229–241.

Hoeg, O. A., and Bose, M. N., 1960, The *Glossopteris* flora of the Belgian Congo with a note on some fossil plants of the Zambesi Basin: Musée Royal du Congo Belge, Sciences Géologiques, Annales, v. 32, p. 1–106.

Holdsworth, B. K., and Nell, P. A. R., 1992, Mesozoic radiolarian faunas from the Antarctic Peninsula: Age, tectonic, and palaeoceanographic significance: Geological Society of London Journal, v. 149, pt. 6, p. 1003–1020.

Hollick, C. A., 1936, The Tertiary Floras of Alaska (with a chapter on the geology of the Tertiary deposits by P. S. Smith): U.S. Geological Survey Professional Paper 182, 185 p.

Honnorez, J., Bonatti, E., Emiliani, C., Bronnimann, P., Furrer, M. A., and Meyerhoff, A. A., 1975, Mesozoic limestone from the Vema offset zone, Mid-Atlantic Ridge: Earth and Planetary Science Letters, v. 26, p. 8–12.

Hou Hongfei and Boucot, A. J., 1990, The Balkhash-Mongolia-Okhotsk Region of the Old World Realm (Devonian): Geological Society of London Memoir 12, p. 297–303.

Hsu Ren, Rigby, J., and Duan Shuying, 1990, Revision of *Glossopteris* flora from southern Xizang: Scientia Geologica Sinica, no. 3, p. 233–242.

Huang Jiqing, 1984, New researches on the tectonic characteristics of China, *in* Tectonics of Asia, Colloquium 05, 27th International Geological Congress, Moscow, 1984, Report 05: Moscow, Izdatelstvo Nauka, p. 13–28.

Huang Jiqing and Chen Bingwei, 1987, The evolution of the Tethys in China and adjacent regions: Beijing, Geological Publishing House, 109 p.

Hu Changming, 1984, Discovery of turbidite from Zhanjin Formation and Qudi Formation in the Late Carboniferous to Early Permian in Duoma district, Rutog, Xizang: Wuhan College of Geology Earth Science Journal, total 25, no. 2, p. 75–78.

Hudson, R. G. S., 1960, The Permian and Trias of the Oman Peninsula, Arabia: Geological Magazine, v. 97, p. 299–308.

Hudson, R. G. S., and Chatton, M., 1959, The Musandam Limestone (Jurassic to Lower Cretaceous) of Oman, Arabia: Museum National d'Histoire Naturelle, Notes et Mémoires sur le Moyen-Orient, t. 7, p. 69–93.

Hudson, R. G. S., and Sudbury, M., 1959, Permian brachiopods from south-east Arabia: Museum National d'Histoire Naturelle, Notes et Mémoires sur le Moyen-Orient, t. 7, p. 19–55.

Ingavat, R., 1984, On the correlation of the Permian foraminiferal faunas of the western, central and eastern provinces of Thailand: Société Géologique de France Mémoire, n.s., no. 147, p. 93–100.

Ingavat, R., and Taquet, P., 1978, First discovery of dinosaur remains in Thailand: Geological Survey of Thailand, Special Issue for GEOSEA III, Geology and Mineral Resources of Thailand, v. 3, pt. 1, p. 1–6.

Isaacson, P. E., and Sablock, P. E., 1988, Devonian System in Bolivia, Peru and northern Chile, *in* McMillan, N. J., Embry, A. F., and Glass, D. J., eds., Devonian of the World, v. I, Regional Syntheses: Canadian Society of Petroleum Geologists Memoir 14, p. 719–728.

Jaeger, J.-J., Adloff, C., Doubinger, J., Pons, D., Vozenin-Serra, C., and Wang Naiwen, 1982, The contribution of fossils to paleogeography of the Lhassa (sic) block (Tibet) [abs.]: Eos (Transactions, American Geophysical Union), v. 63, p. 1093–1094.

Jain, S. L., 1980, The continental Lower Jurassic fauna from the Kota Formation, India, *in* Jacobs, L. L., ed., Aspects of Vertebrate History: Flagstaff, Museum of Northern Arizona, p. 99–123.

Jain, S. L., 1983, A review of the genus *Lepidotes* (Actinopterygii: Semionotoformes), with special reference to the species from the Kota Formation (Lower Jurassic), India: Palaeontological Society of India Journal, v. 28, p. 7–42.

Jardine, N., and McKenzie, D., 1972, Continental drift and the dispersal and evolution of organisms: Nature, v. 235, p. 20–24.

Jardine, S., Micholet, J., Molinas, E., and Penet, B., 1969, Présence de Permien au Gabon: identification stratigraphique et sédimentologique du Groupe de l'Agoula: Compte Rendu Sommaire, Séances, Société Géologique de France, fasc. 8, p. 297–298.

Jarvik, E., 1961, Devonian vertebrates, *in* Raasch, G. O., ed., Geology of the Arctic, 1: Toronto, University Toronto Press, p. 197–204.

Jenkins, T. B. H., and Jenkins, D. G., 1971, Conodonts of the Haast Schist and Torlesse Groups of New Zealand. Part 1-Biostratigraphic significance of the Triassic conodonts from the Mount Mason and Okuku Limestones: New Zealand Journal of Science and Technology, v. 14, p. 782–794.

Jepson, W. L., 1925, A Manual of the Flowering Plants of California: San Francisco, California School Book Depository, 1238 p.

Jin Yugan, 1981, On the paleoecological relation between Gondwana and Tethys faunas in the Permian of Xizang, *in* Liu Dongsheng, ed., Proceedings, Symposium on Qinghai-Xizang (Tibet) Plateau, Beijing, Geological and Ecological Studies of Qinghai-Xizang Plateau, v. 1, Geology. Geological History and Origin of Qinghai-Xizang Plateau, Beijing: New York, Science Press, and Gordon and Breach, Science Publishers, p. 171–178.

Jin Yugan, 1985 [1986], Permian Brachiopoda and paleogeography of the Qinghai-Xizang (Tibet) Plateau: Palaeontologia Cathayana, no. 2, p. 19–71.

Johnson, R. D., Powell, C. McA., and Veevers, J. J., 1976, Spreading history of the eastern Indian Ocean and Greater India's northward flight from Antarctica and Australia: Geological Society of America Bulletin, v. 87, p. 1560–1566.

Jongmans, W. J., 1937, The flora of the Upper Carboniferous of Djambi (Sumatra, Netherlands Indies) and its possible bearing on the palaeogeography of the Carboniferous: Congrès pour l'Avancement des Études de Stratigraphie Carbonifère, 2nd, Heerlen 1935: Compte Rendu, v. 1, p. 327–344.

Jongmans, W. J., 1940, Beitrage zur Kenntniss der Karbonflora von niederlandisch Neu-Guinea: Netherlands, Geologisch Bureau voor het Mijngebied te Heerlen, Jaarverslag 1938–39, Mededeelingen, p. 263–274.

Just, T. K., 1952, Fossil floras of the Southern Hemisphere and their phytogeographic significance: Bulletin of the American Museum of Natural History, v. 99, p. 189–202.

Kalandadze, N. N., 1974, O mezhkontinental'nykh svyazyakh faun tetrapod v triasovom periode: Paleontologicheskii Zhurnal, no. 3, p. 75–86.

Kalandadze, N. N., 1975, Pervaya nakhoda lystrozavra na territorii evropeyskoy chasti SSSR: Paleontologicheskii Zhurnal, no. 4, p. 140–142.

Kalandadze, N. N., and Rautian, A. S., 1983, Mesto Tsentralnoy Azii v zoogeograficheskoy istorii Mesozoya, *in* Barsbold, R., Vorob'yeva, E.I., Lubsandanzan, B., Tatarinov, L. P., Trofimov, B. A., Reshchetov, V. Yu., and Shishchkin, M. A., eds., Iskopayemyye reptilii Mongolii: Moscow, Nauka, Sovmestnaya Sovetsko-Mongol'skaya Paleontologicheskaya Ekspeditsiya, Trudy, v. 24, p. 6–44.

Kamen-Kaye, M., 1972, Permian Tethys and Indian Ocean: American Association of Petroleum Geologists Bulletin, v. 56, p. 1984–1999.

Kamen-Kaye, M., 1978, Permian to Tertiary faunas and paleogeography: Somalia, Kenya, Tanzania, Mozambique, Madagascar, South Africa: Journal of Petroleum Geology, v. 1, p. 79–101.

Kamen-Kaye, M., 1985, Mesozoic columns below the Seychelles Bank, western Indian Ocean: Journal of Petroleum Geology, v. 8, p. 323–329.

Kamen-Kaye, M., and Meyerhoff, A. A., 1979, Flood basalts: implications for global-tectonic hypotheses and petroleum exploration: Journal of Petroleum Geology, v. 1, p. 29–37.

Kamen-Kaye, M., and Meyerhoff, A. A., 1980, Petroleum geology of the Mascarene Ridge, western Indian Ocean: Journal of Petroleum Geology, v. 3, p. 123–138.

Kapoor, H. M., 1979, Gondwana of Kashmir—A reappraisal, *in* Laskar, B., and Raja Rao, C. S., eds., 4th International Gondwana Symposium, Calcutta, 1977: Delhi, Hindustan Publishing Corporation, v. 2, p. 443–462.

Kapoor, H. M., and Tokuoka, T., 1985, Sedimentary facies of the Permian and Triassic of the Himalayas, *in* Nakazawa, K., and Dickins, J. M., eds., The Tethys: Tokyo, Tokai University Press, p. 23–58.

Kashfi, M. S., 1976, Plate tectonics and structural evolution of the Zagros geosyncline, southwestern Iran: Geological Society of America Bulletin, v. 87, p. 1486–1490.

Kauffman, E., 1987, Our uniformitarian albatross: Palaios, v. 2, p. 531.

Keeley, M. L., Dungworth, G., Floyd, C. S., Forbes, G. A., King, C., McGarve, R. M., and Shaw, D., 1990, The Jurassic system in northern Egypt: Journal of Petroleum Geology, v. 13, p. 397–420.

Kemp, E. M., Balmé, B. E., Helby, R. J., Kyle, R. A., Playford, G., and Price, P. E., 1977, Carboniferous and Permian palyno-stratigraphy in Australia and Antarctica: A review: Bureau of Mineral Resources, Journal of Australian Geology and Geophysics, v. 2, p. 177–208.

Kent, P. E., Hunt, J. A., and Johnstone, D. W., 1971, The geology and geophysics of coastal Tanzania: London, Institute of Geological Sciences, Natural Environment Research Council, Geophysical Paper no. 6, 101 p.

Keyes, I. W., 1977, Records of the northern hemisphere Cretaceous genus *Onchopristus* (Order Batoidea) from New Zealand: New Zealand Journal of Geology and Geophysics, v. 20, p. 263–272.

Keyser, A. W., 1981, The stratigraphic distribution of the *Dicynodontia* of Africa reviewed in a Gondwana context, *in* Cresswell, M. M., and Vella, P., eds., Gondwana Five: Rotterdam, A. A. Balkema, p. 61–63.

Khudoley, K. M., 1974, Circum-Pacific Mesozoic ammonoid distribution: relation to hypotheses of continental drift, polar wandering, and earth expansion, *in* Kahle, C. F., ed., Plate Tectonics—Assessments and Reassessments: American Association of Petroleum Geologists Memoir 23, p. 295–330.

Khudoley, K. M., 1988, The paleobiogeography of the Atlantic Ocean in the Jurassic Period: International Geology Review, v. 30, p. 623–634.

Khudoley, K. M., and Prosorovskaya, E.L., 1985, Seaways between the western part of the Pacific and adjacent oceans, *in* Westermann, G. E. G., ed., Jurassic Biogeography and Stratigraphy of East U.S.S.R.: International Geological Congress Project, no. 171, Circum-Pacific Jurassic, Special Paper 10, 29 p.

King, P. B., 1951, The Tectonics of Middle North America. Middle North America East of the Cordilleran System: Princeton, New Jersey, Princeton University Press, 203 p.

Kitching, J. W., Collins, J. W., Elliot, D. H., and Colbert, E. H., 1972, *Lystrosaurus* Zone (Triassic) fauna from Antarctica: Science, v. 175, p. 524–527.

Klinger, H. C., Kennedy, W. J., and Dingle, R. V., 1972, A Jurassic ammonite from South Africa: Neues Jahrbuch für Geologie und Paläontologie, Monatshefte 1972, no. 11, p. 653–659.

Klitzsch, E., and Lejal-Nicol, A., 1984, Flora and fauna from strata in southern Egypt and northern Sudan (Nubia and surrounding areas): Berliner Geowissenschaft, Abhandlungen, reihe A, bd. 50, p. 47–79.

Kobayashi, T., and Hamada, T., 1980, Carboniferous trilobites of Japan: Paleontological Society of Japan, Special Paper 23, 132 p.

Koken, E., 1907, Indisches Perm und die permische Eiszeit: Neues Jahrbuch für Mineralogie, Geologie und Paläontologie, festband, p. 446–545.

Kon'no, E., 1963, Some Permian plants from Thailand: Japanese Journal of Geology and Geography, v. 34, p. 139–159.

Kon'no, E., 1968, The Upper Permian flora from the eastern border of northeast China: Tohoku University Science Reports, ser. 2, Geology, v. 39, p. 159–211.

Koshal, V. N., 1984, Differentiation of Rhaetic sediments in sub-surface of Kutch based on palynofossils, *in* Petroliferous basins of India—II: Dehra Dun, Petroleum Asia Journal, v. 7, p. 102–105.

Kovacs-Endrody, E., 1991, On the Late Permian age of Ecca *Glossopteris* floras in the Transvaal Province with a key to and description of twenty five *Glossopteris* species: Memoirs of the Geological Survey of South Africa, v. 77, 111 p.

Krassilov, V. A., 1987, Palaeobotany of the Mesophyticum: State of the art: Review of Palaeobotany and Palynology, v. 50, p. 231–254.

Kruck, W., and Thiele, J., 1983, Late Paleozoic glacial deposits in the Yemen Arab Republic: Geologische Jahrbuch, reihe B, bd. 46, p. 3–29.

Kumar, G., Bhatt, D. K., and Raina, B. K., 1987, Skeletal microfauna of Meishucunian and Qiongzhusian (Precambrian-Cambrian boundary) age from the Ganga Valley, Lesser Himalaya: Geological Magazine, v. 124, p. 167–171.

Kumar, S. P., 1983, Krishna-Godavari-Cauvery, *in* Petroliferous Basins of India: Dehra Dun, Petroleum Asia Journal, v. 6, no. 4, p. 57–65.

Kummel, B., 1950, Stratigraphic studies in northern Peru: American Journal of Science, v. 248, p. 249–263.

Kummel, B., 1969, Ammonoids of the late Scythian (Lower Triassic): Harvard Museum of Comparative Zoology Bulletin, v. 137, p. 311–701.

Kummel, B., and Teichert, C., eds., 1970a, Stratigraphic Boundary Problems: Permian and Triassic of West Pakistan: University Press of Kansas Department of Geology Special Publication 4, 474 p.

Kummel, B., and Teichert, C., 1970b, Stratigraphy and paleontology of the Permian-Triassic boundary beds, Salt Range and Trans-Indus Ranges, West Pakistan, *in* Kummel, B., and Teichert, C., eds., Stratigraphic Boundary Problems: Permian and Triassic of West Pakistan: University Press of Kansas Department of Geology Special Publication 4, p. 2–110.

Kutty, T. S., 1972, Permian reptilian fauna from India: Nature, v. 237, p. 462–463.

Lacey, W. S., 1975, Some problems of "mixed" floras in the Permian of Gondwanaland, *in* Campbell, K. S. W., ed., Gondwana Geology: Canberra, Australian National University Press, p. 125–134.

Lacey, W. S., and Huard-Moine, D., 1966, Karroo floras of Rhodesia and Malawi, Pt 2. The *Glossopteris* flora in the Wankie district of Southern Rhodesia, *in* Symposium, Floristic Stratigraphy of Gondwanaland: Lucknow, India, Birbal Sahni Institute, p. 13–25.

Lacey, W. S., and Smith, C. S., 1972, Studies on Karoo floras from the Upper Luangwa Valley, Zambia, *in* 2nd Gondwana Symposium, South Africa, July to August, 1970, Proceedings and Papers: Pretoria, Council for Scientific and Industrial Research, p. 571–574.

Lane, N. G., and Frakes, L. A., 1970, A Permian starfish from South West Africa: Journal of Paleontology, v. 44, p. 1135–1136.

Legrand-Blain, M., 1985a [1987a], Egypt, *in* Martinez-Diaz, C., ed., North Africa, The Carboniferous of the world, IUGS Publication 20, II—Australia, Indian Subcontinent, South Africa, South America, and North Africa: Madrid, Instituto Geológico y Minero de España, p. 347–351.

Legrand-Blain, M., 1985b [1987b], Brachiopods, *in* Martinez-Diaz, C., ed., North Africa, The Carboniferous of the World, IUGS Publication 20, II—Australia, Indian Subcontinent, South Africa, South America, and North Africa: Madrid, Instituto Geológico y Minero de España, p. 172–174.

Leip, H., 1958, The River in the Sea: New York, G. P. Putnam's Sons, 223 p.

Lejal-Nicol, A., 1985 [1987], Microfloras, *in* Martinez-Diaz, C., ed., North Africa, The Carboniferous of the world, IUGS Publication. 20, II—Australia, Indian Subcontinent, South Africa, South America, and North Africa: Madrid, Instituto Geológico y Minero de España, p. 386–391.

Liang Dingye, Nie Zetong, Guo Tieying, Xu Baowen, Zhang Yizhe, and Wang Weipin, 1983, Permo-Carboniferous Gondwana-Tethys facies in southern Karakoran [sic] Ali, Xizang (Tibet): Wuhan College of Geology Earth Science Journal, total 19, no. 1, p. 9–27.

Liu Benpei, and Cui Xinsheng, 1983, Discovery of *Eurydesma*-fauna from Rutog, northwest Xizang (Tibet): Wuhan College of Geology Earth Science Journal, total 19, no. 1, p. 79–92.

Liu Benpei, Feng Qinglai, and Fang Nianqiao, 1991, Tectonic evolution of the Palaeo-Tethys in Changming-Menglian belt and adjacent regions, western Yunnan: Journal of China University of Geosciences, v. 2, p. 18–28.

Li Xingxue, and Wu Xiuyuan, 1993, The Cathaysian and Gondwana floras: Their contributions to determining the boundary between eastern Gondwana and Laurasia: Journal of Southeast Asian Earth Sciences, v. 9, p. 309–317.

Loczy, L. de, 1970, Role of transcurrent faulting in South American tectonic framework: American Association of Petroleum Geologists Bulletin, v. 54, p. 2111–2119.

Loock, J. C., and Visser, N. J., 1985 [1987], South Africa, *in* Martinez-Diaz, C., ed., South Africa, The Carboniferous of the World, IUGS Publication 20, II—Australia, Indian Subcontinent, South Africa, South America, and North Africa: Madrid, Instituto Geológico y Minero de España, p. 167–174.

Lopez-Gamundi, O. R., Cesari, S. N., and Limarino, C.O., 1993, Paleoclimatic significance and age constraints of the Carboniferous coals of Paganzo basin, western Argentina, *in* Findlay, R. H., Unrug, R., Banks, M. R., and Veevers, J. J., eds., Gondwana Eight: Rotterdam, A. A. Balkema, p. 291–298.

Loughnan, F. C., 1975, Laterites and flint clays in the Early Permian of the Sydney basin, Australia, and their paleoclimatic implications: Journal of Sedimentary Petrology, v. 45, p. 591–598.

Lowman, P. D., Jr., 1985, Plate tectonics with fixed continents: A testable hypothesis—I: Journal of Petroleum Geology, v. 8, p. 373–378.

Lowman, P. D., Jr., 1986, Plate tectonics with fixed continents: A testable hypothesis—II: Journal of Petroleum Geology, v. 9, p. 71–87.

Lys, M., 1985 [1987], Foraminifera, *in* Martinez-Diaz, C., ed., North Africa, The Carboniferous of the world, IUGS Publication 20, II—Australia, Indian Subcontinent, South Africa, South America, and North Africa: Madrid, Instituto Geológico y Minero de España, p. 354–364.

Machens, E., 1973, The geological history of the marginal basins along the north shore of the Gulf of Guinea, *in* Nairn, A. E. M., and Stehli, F. G., eds., The ocean basins and margins, v. 1, The South Atlantic: New York, Plenum Press, p. 351–390.

Maithy, P. K., 1976 [1978], Further observations on Indian Lower Gondwana Sphenophyllales: The Palaeobotanist, v. 25, p. 266–278.

Mamet, B., 1972, Considerations paléogéographiques deduites de l'étude des foraminifères des couches de passage de Viséen au Namurien (Bassins de Reggane et de Fort-Polignac, Sahara Central): Institut Royal des Sciences Naturelles de Belgique, Sciences de la Terre, Bulletin 48, no. 8, p. 1–13.

Mamet, B., 1973, Foraminiferal zonation of the Carboniferous: Stratigraphic implications: Geological Society of America, Abstracts with Programs, v. 5, p. 725.

Mamet, B. L., and Belford, D. J., 1968, Carboniferous foraminifera, Bonaparte Gulf basin, northwestern Australia: Micropaleontology, v. 14, p. 339–347.

Mamet, B. L., and Playford, P. E., 1968, Sur la présence de Quasiendothyrinae (foraminifères) en Australie occidentale: Société Géologique de France, Compte Rendu, no. 7, p. 229–230.

Marques Toigo, M., 1991, Palynobiostratigraphy of the southern Neopaleozoic Gondwana sequence, *in* Ulbrich, H., and Rocha Campos, A. C., eds., Gondwana Seven Proceedings: Sao Paulo, Instituto de Geociencias, Universidade do Sao Paulo, p. 503–515.

Martin, G., 1982, Geologie des Kustengebietes von Nordwest-Afrika südlich der Sahara. Neue Erkenntnisse aus der Erdolexploration: Lenz, Giessner Geologisches Schriften no. 30, 150 p.

Martin, H., 1953, Notes on the Dwyka succession and some pre-Dwyka valleys in South West Africa: Transactions of the Geological Society of South Africa, v. 56, p. 37–41.

Martin, H., 1973, The Atlantic margin of southern Africa between latitude 17° South and the Cape of Good Hope, *in* Nairn, A. E. M., and Stehli, F. G., eds., The Ocean Basins and Margins. 1. The South Atlantic: New York, Plenum Press, p. 277–300.

Martin, H., 1975, Structural and palaeogeographic evidence for an upper Palaeozoic sea between southern Africa and South America, *in* Campbell, K. S. W., ed., Gondwana Geology: Canberra, Australian National University Press, p. 37–51.

Martin, H., Walliser, O. H., and Wilczewski, N., 1971, A goniatite from the glaciomarine Dwyka beds near Schlip, South West Africa, *in* Haughton, S. H., ed., Proceedings and Papers of 2nd Gondwana Symposium, South Africa 1970: Pretoria, Geological Society of South Africa, p. 621–625.

Massa, D., Havlicek, V., and Bonnefous, J., 1977, Stratigraphic and faunal data on the Ordovician of the Rhadames Basin (Libya and Tunisia): Bulletin du centre de Recherche et Exploration-Production Elf-Aquitaine, v. 1, p. 3–27.

Matsuda, T., 1985, Late Permian to Early Triassic conodont paleobiogeography in the "Tethys Realm," *in* Nakazawa, K., and Dickins, J. M., eds., The Tethys: Tokyo, Tokai University Press, p. 157–170.

McClure, H. A., 1980, Permian-Carboniferous glaciation in the Arabian Peninsula: Geological Society of America Bulletin, v. 91, p. 707–712.

McClure, H., and Young, G. M., 1981, Late Palaeozoic glaciation in the Arabian Peninsula, *in* Hambrey, M. J., and Harland, W. B., eds., Earth's Pre-Pleistocene Glacial Record: Cambridge, Cambridge University Press, p. 275–277.

McGowan, J. A., 1971, Oceanic biogeography of the Pacific, *in* Funnell, B. M., and Riedel, W. R., eds., The Micropaleontology of the Oceans: Cambridge, Cambridge University Press, p. 3–74.

McLachlan, I. R., and Anderson, A. M., 1973, A review of the evidence for marine conditions in southern Africa during Dwyka times: Palaeontographica Africana, v. 15, p. 37–64.

McLachlan, I. R., and Anderson, A. M., 1975, The age and stratigraphic relationship of the glacial sediments in southern Africa, *in* Campbell, K. S. W., Gondwana Geology: Canberra, Australian National University Press, p. 415–422.

McTavish, R. A., 1975, Triassic conodonts and Gondwana stratigraphy, *in* Campbell, K. S. W., ed., Gondwana Geology: Canberra, Australian National University Press, p. 481–490.

McWhae, J. R. H., Playford, P. E., Lindner, A. W., Glenister, B. F., and Balme, B. E., 1958, The stratigraphy of Western Australia: Journal of the Geological Society of Australia, v. 4, pt. 2, 161 p.

Mehra, S., Verma, K. K., and Srivastava, S., 1979, Contribution to the study of Upper Gondwanas of Kutch, Gujarat, India, *in* Laskar, B., and Raja Rao, C.S., eds., 4th International Gondwana Symposium, Calcutta, 1977: Dehli, Hindustan Publishing Corporation, v. 2, p. 491–495.

Melo, J. H. G. de, 1988, The Malvinokaffric Realm in the Devonian of Brazil, *in* McMillan, N. J., Embry, A. F., and Glass, D. J., eds., Devonian of the World, v. I, Regional Syntheses, Canadian Society of Petroleum Geologists Memoir 14, p. 669–704.

Mercer, J. H., 1983, Cenozoic glaciation in the Southern Hemisphere: Annual Review of Earth and Planetary Sciences, v. 11, p. 99–132.

Mesigos, M., 1953, El Paleozóico superior de Barreal y su continuación austral "Sierra de Barreal" (Prov. San Juan): Asociación Geológicc de Argentina, Revista, 8, no. 2, p. 65–109.

Metcalfe, I., 1981, Permian and Early Triassic conodonts from northwest peninsular Malaysia: Geological Society of Malaysia Bulletin, no. 14, p. 119–126.

Metcalfe, I., 1984, Stratigraphy, palaeontology and palaeogeography of the Carboniferous of Southeast Asia: Société Géologique de France Mémoire, n.s., no. 147, p. 107–118.

Metcalfe, I., 1986, Late Palaeozoic palaeogeography of southeast Asia: Some stratigraphical, palaeontological and palaeomagnetic constraints: Geological Society of Malaysia Bulletin, no. 19, p. 153–164.

Metcalfe, I., 1990, Triassic conodont biostratigraphy in the Malay Peninsula: Geological Society of Malaysia Bulletin, no. 26, p. 133–145.

Metcalfe, I., Idris, M., and Tan, J. T., 1980, Stratigraphy and palaeontology of the Carboniferous sediments in the Panching area, Pahang, West Malaysia: Geological Society of Malaysia Bulletin, no. 13, p. 1–26.

Meyen, S. V., 1969, New data on relationship between Angara and Gondwana late Palaeaozoic floras, *in* Gondwana Stratigraphy, IUGS Symposium, Mar del Plata [*non* Buenos Aires], 1-15 October 1967; Paris, United Nations Educational, Scientific and Cultural Organization (UNESCO), Earth Sciences 2, p. 141–157.

Meyerhoff, A. A., 1952, A study of leaf venation in the Betulaceae, with its application to paleobotany [Ph.D. thesis]: Palo Alto, California, Stanford University, 248 p.

Meyerhoff, A. A., 1970a, Continental drift: Implications of paleomagnetic studies, meteorology, physical oceanography, and climatology: Journal of Geology, v. 78, p. 1–51.

Meyerhoff, A. A., 1970b, Continental drift, II: High-latitude evaporite deposits and geologic history of Arctic and North Atlantic oceans: Journal of Geology, v. 78, p. 406–444.

Meyerhoff, A. A., 1984, Carboniferous oil and gas generation in the Eastern Hemisphere—II: Journal of Petroleum Geology, v. 7, p. 313–328.

Meyerhoff, A. A., and Meyerhoff, H. A., 1974, Tests of plate tectonics, *in* Kahle, C. F., ed., Plate tectonics—Assessments and reassessments: American Association of Petroleum Geologists Memoir 23, p. 43–145.

Meyerhoff, H. A., and Meyerhoff, A. A., 1978, Spreading history of the eastern Indian Ocean and India's northward flight from Antarctica and Australia: Discussion: Geological Society of America Bulletin, v. 89, p. 637–640.

Meyerhoff, A. A., and Teichert, C., 1971, Continental drift, III: Late Paleozoic glacial centers, and Devonian-Eocene coal distribution: Journal of Geology, v. 79, p. 285–321.

Meyerhoff, A. A., Taner, I., and Kamen-Kaye, M., 1987, Letters: Geology: v. 15, p. 95–96.

Meyerhoff, A. A., Kamen-Kaye, M., Chen Chin, and Taner, I., 1991, China: Stratigraphy, paleogeography and tectonics: Dordrecht, Kluwer Academic Publishers, 188 p.

Meyerhoff, A. A., Taner, I., Morris, A. E. L., Martin, B. D., Jr., Agocs, W. B., and Meyerhoff, H. A., 1992, Surge tectonics: a new hypothesis of earth dynamics: *in* Chatterjee, S., and Hotton, N., III, eds., New Concepts in Global Tectonics: Lubbock, Texas Tech University Press, p. 309–409.

Mikhaylov, Y. A., Ustritskiy, V. I., Chernyak, G. Y., and Yavshits, G. P., 1970, Upper Permian glaciomarine sediments of the northeastern USSR: Akademia Nauk SSSR, Doklady (English trans.), American Geological Institute, v. 190, p. 100–102.

Mildenhall, D. C., 1970, Discovery of a New Zealand member of the Permian *Glossopteris* flora: Australian Journal of Science, v. 32, p. 474–475.

Millot, J., 1972, In Conclusion, *in* Battistini, R., and Richards-Vindard, G., eds., Biogeography and Ecology in Madagascar: The Hague, W. Junk, p. 741–756.

Milner, A. R., 1993, Biogeography of Paleozoic tetrapods, *in* Long, J. A., ed., Paleozoic Vertebrate Biostratigraphy and Biogeography: Baltimore, Johns Hopkins University Press, p. 324–353.

Minato, M., and Kato, M., 1977, Tethys sea corals in the upper Palaeozoic: Bureau de Recherches Géologiques et Minières, Mémoires, Paris, no. 89, p. 228–233.

Molnar, R. E., 1981, A dinosaur from New Zealand, *in* Cresswell, M. M., and Vella, P., eds. Gondwana Five: Rotterdam, A. A. Balkema, p. 91–96.

Molnar, R. E., 1992, Paleozoogeographic relationships of Australian Mesozoic tetrapods, *in* Chatterjee, S., and Hotton, N., III, eds., New Concepts in Global Tectonics: Lubbock, Texas Tech University Press, p. 259–266.

Morris, N., 1975, The *Rhacopteris* flora in New South Wales, *in* Campbell, K. S. W., ed., Gondwana geology: Canberra, Australian National University Press, p. 99–108.

Mu En-zhi, Wen Shixuan, Wang Yigang, Zhang Bingkao, and Jin Jixiang, 1973, Stratigraphy of the Mount Jolmo Lungma region in southern Tibet, China: Scientia Sinica, v. 16, no. 1, p. 96–111.

Mu En-zhi, Boucot, A. J., Chen Xu, and Rong Jiayu, 1986, Correlation of the Silurian rocks of China: Geological Society of America Special Paper 202, 80 p.

Murris, R. J., 1980, Hydrocarbon habitat of the Middle East, *in* Miall, A. D., ed., Facts and principles of world petroleum occurrence: Canadian Society of Petroleum Geologists Memoir 6, p. 765–800.

Murthy, N. G. K., and Ahmad, M., 1979, Paleogeographic significance of the Talchirs in the Palar basin near Madras, India, *in* Laskar, B., and Raja Rao, C. S., eds., 4th Internatational Gondwana Symposium, Calcutta, 1977: Delhi, Hindustan Publishing Corporation, v. 2, p. 515–521.

Nakamura, K., Shimizu, D., and Liao Zhuoting, 1985, Permian palaeobiogeography of brachiopods based on the faunal provinces, *in* Nakazawa, K., and Dickins, J. M., eds., The Tethys: Tokyo, Tokai University Press, p. 185–198.

Nakazawa, K., 1985, The Permian and Triassic Systems in the Tethys—Their paleogeography, *in* Nakazawa, K., and Dickins, J. M., eds., The Tethys: Tokyo, Tokai University Press, p. 93–111.

Nakazawa, K., and Dickins, J. M., eds., 1985, The Tethys: Tokyo, Tokai University Press, 317 p.

Nakazawa, K., and Kapoor, H. M., 1979, Correlation of the marine Permian in the Tethys and Gondwana, *in* Laskar, B., and Raja Rao, C. S., eds., 4th International Gondwana Symposium, Calcutta, 1979: Delhi, Hindustan Publishing Corporation, v. 2, p. 409–419.

Nakazawa, K., and Kapoor, eds., 1981, The Upper Permian and Lower Triassic faunas of Kashmir: Geological Survey of India Memoirs, Palaeontologia Indica, n.s., v. 46, 205 p.

Nakazawa, K., Kapoor, H. M., Ishii, K., Bando, Y., Okimura, Y., and Tokuoka,

T., 1975, The Upper Permian and the Lower Triassic in Kashmir, India: Kyoto University Faculty of Science Memoirs, Series of Geology and Mineralogy, v. 42, p. 1–106.

Neaverson, E., 1955, Stratigraphic palaeontology. A study of ancient life-provinces, (second edition): Oxford, Clarendon Press, 806 p.

Neuburg, M. F., 1948, Verkhnepaleozoyskaya flora Kuznetskogo basseyna: Akademia Nauk SSSR, Paleontologiya SSSR, v. 12, pt. 3, fasc. 2, 342 p.

Newell, N. D., 1949, Geology of the Lake Titicaca Region, Peru and Bolivia: Geological Society of America Memoir 36, 111 p.

Newell, N. D., Chronic, J., and Roberts, T. G., 1953, Upper Paleozoic of Peru: Geological Society of America Memoir 58, 279 p.

Noetling, F., 1901, Beiträge zur Geologie der Salt Range: Neues Jahrbuch für Mineralogie, Geologie und Paläontologie, v. 14, p. 369–471.

Norin, E., 1930, An occurrence of late Palaeozoic tillite in the Kuruk-tagh Mountains, central Asia: XV International Geological Congress, South Africa 1929, Compte Rendu, v. II: Pretoria, Wallachs' Ltd. Printers, p. 74–76.

Oelofsen, B. W., 1987, The biostratigraphy and fossils of the Whitehill and Irati shale formations of the Karoo and Parana basins, *in* McKenzie, G. D., ed., Gondwana Six: Stratigraphy, Sedimentology, and Paleontology: American Geophysical Union, Geophysical Monograph 41, p. 131–138.

Oesterlen, P. M., 1990, A new occurrence of the Dwyka Formation in western Zimbabwe: South African Journal of Geology, v. 93, p. 382–388.

Ojakangas, R. W., and Matsch, C. L., 1981, The late Palaeozoic Whiteout Conglomerate: A glacial and glaciomarine sequence in the Ellsworth Mountains, West Antarctica, *in* Hambrey, M. J., and Harland, W. B., eds., Earth's Pre-Pleistocene Glacial Record: Cambridge, Cambridge University Press, p. 241–244.

Okimura, Y., Ishii, K., and Ross, C. A., 1985, Biostratigraphical significance and faunal provinces of Tethyan Late Permian smaller foraminifera, *in* Nakazawa, K., and Dickins, J. M., eds., The Tethys: Tokyo, Tokai University Press, p. 115–138.

Oviedo, G.C., 1965, Estratigrafía de la Peninsula de Copacabana, Lago Titicaca, Departamento de La Paz: Instituto Boliviano del Petroleo Boletin, v. 5, p. 5–15.

Page, B. G. N., 1981, Late Palaeozoic pebbly mudstones in Sumatra, *in* Hambry, M. J., and Harland, W. B., eds., Earth's Pre-Pleistocene Glacial Record: Cambridge, Cambridge University Press, p. 337.

Pakistani-Japanese Research Group, 1985, Permian and Triassic Systems in the Salt Range and Surghar Range, *in* Nakazawa, K., and Dickins, J. M., eds., The Tethys: Tokyo, Tokai University Press, p. 221–312.

Pal, A. K., and Chaloner, W. G., 1979, A Lower Carboniferous *Lepidodendropsis* flora in Kashmir: Nature, v. 281, p. 295–297.

Pantic, N., Hochuli, P. A., and Gansser, A., 1981, Jurassic palynomorphs beneath the Main Central Thrust of east Bhutan (Himalayas): Eclogae Geologicae Helvetiae, v. 74, p. 883–892.

Parker, R. A., Houghton, H. F., and McDowell, R. C., 1988, Stratigraphic framework and distribution of early Mesozoic rocks of the northern Newark basin, New Jersey and New York, *in* Froelich, A. J., and Robinson, G. R., Jr., eds., Studies of the Early Mesozoic Basins of the Eastern United States: U.S. Geological Survey Bulletin 1776, p. 31–39.

Pascoe, E.H., 1959, A Manual of the Geology of India and Burma, v. II, (third edition): Calcutta, Government of India Press, 1343 p.

Patterson, C., and Owen, H. G., 1991, Indian isolation or contact? A response to Briggs: Systematic Zoology, v. 40, p. 96–100.

Paul, G. S., 1988, Physiological, migratorial, climatological, geophysical, survival, and evolutionary implications of Cretaceous polar dinosaurs: Journal of Paleontology, v. 62, p. 640–652.

Petri, S., and Fulfaro, V. J., 1983, Geología do Brasil: Sao Paulo, Editora da Universidade Sao Paulo, 631 p.

Pickett, J. W., and Wu, W., 1990, The succession of Early Carboniferous coral faunas in eastern Australia and China: Alcheringa, v. 14, no. 2, p. 89–108.

Plumstead, E. P., 1973, The late Palaeozoic *Glossopteris* flora, *in* Hallam, A., ed., Atlas of Palaeobiogeography: Amsterdam, Elsevier Publishing Co., p. 187–205.

Pogue, K. R., Wardlaw, B. R., Harris, A. G., and Hussain, A., 1992, Paleozoic and Mesozoic stratigraphy of the Peshawar basin, Pakistan: Correlations and implications: Geological Society of America Bulletin, v. 104, p. 915–927.

Pons, P., and Vozenin-Serra, C., 1984, Les bois aptiens de la région de Lhassa (Sud-Xizang, Tibet): Implications paléogéographiques, paléoclimatiques et paléoecologiques: Société Géologique de France Mémoire, n.s., no. 147, p. 118–124.

Pszczolkowski, A., 1989, Late Paleozoic fossils from pebbles in the San Cayetano Formation, Sierra del Rosario, Cuba: Annales Societatis Geologorum Poloniae, v. 51, p. 27–40.

Pudsey, C. J., Schroeder, R., and Skelton, P. W., 1986, Cretaceous (Aptian/Albian) age for island-arc volcanics, Kohistan, N. Pakistan, *in* Gupta, V. J., ed., Geology of Western Himalayas, Contributions to Himalayan Geology, v. 3: Delhi, Hindustan Publishing Corporation, p. 150–168.

Quilty, P. G., 1970, Jurassic ammonites from Ellsworth Land, Antarctica: Journal of Paleontology, v. 44, p. 110–116.

Quilty, P. G., 1977, Late Jurassic bivalves from Ellsworth Land, Antarctica: Their systematics and paleogeographic implications: New Zealand Journal of Geology and Geophysics, v. 20, p. 1033–1080.

Quilty, P. G., 1982, Tectonic and other implications of Middle-Upper Jurassic rocks and marine faunas from Ellsworth Land, Antarctica, *in* Craddock, C., ed., Antarctic Geoscience: International Union of Geological Sciences, ser. B, no. 4, University of Wisconsin Press, p. 669–678.

Quilty, P. G., 1983, Bajocian bivalves from Ellsworth Land, Antarctica: New Zealand Journal of Geology and Geophysics, v. 26, p. 395–418.

Racheboeuf, P. R., Boucot, A. J., and Saul, J. M., 1989, A Lower Carboniferous brachiopod fauna from the Sekondi area, coastal Ghana: Neues Jahrbuch für Geologie und Paläontologie, Monatshefte, p. 223–232.

Raju, D. S. N., 1968, Preliminary study on the Cretaceous planktonic foraminifera from the subsurface section in Cauvery basin, south India [abs.]: Geologische Bundesanstalt Verhandlungen, no. 1–3, p. A91–A92.

Ramanthan, S., 1968, Stratigraphy of Cauvery basin with reference to its oil prospects, *in* Cretaceous-Tertiary formations of south India: Geological Society of India, Memoir 2, p. 153–167.

Ranga Rao, Dhar, C. L., and Obergfell, F. A., 1979, Badhaura Formation of Rajasthan—Its stratigraphy and age, *in* Laskar, B., and Raja Rao, C. S., eds., 4th International Gondwana Symposium, Calcutta, 1977: Delhi, Hindustan Publishing Corporation, p. 481–488.

Rasheed, D. A., 1978, Occurrence of *Patellina subcretacea* in Cullygoody (Dalmiapuram) Limestone, Trichinopoly Cretaceous, Tamil Nadu State, India: Delhi, Hindustan Publishing Corporation, Recent Researches in Geology 4, p. 470–473.

Reed, F. R. C., 1928, A Permo-Carboniferous marine fauna from the Umaria coal-field: Geological Survey of India, Records, v. 60, p. 267–298.

Retallack, G. J., 1987, Triassic vegetation and geography of New Zealand portion of the Gondwana supercontinent, *in* McKenzie, G. D., ed., Gondwana Six, Stratigraphy, Sedimentology, and Paleontology: American Geophysical Union Geophysical Monograph 41, p. 29–39.

Richter, M., and Thomson, M. R. A., 1989, First Aspidorhynchidae (Pisces: Teleostei) from Antarctica: Antarctic Science, v. 1, p. 57–64.

Richter, R., 1941, Devon: Geologische Jahresberichte, v. 3A, p. 31–43.

Roberts, J., 1985, Brachiopods, *in* Roberts, J., ed., Australia, *in* Martinez-Diaz, C., ed., The Carboniferous of the World, IUGS Publication 20, II—Australia, Indian Subcontinent, South Africa, South America, and North Africa: Madrid, Instituto Geológico y Minero de España, p. 105–108.

Roberts, J., and Jell, P. A., 1990, Early Middle Cambrian (Ordian) brachiopods of the Coonigan Formation, western New South Wales: Alcheringa, v. 14, p. 257–309.

Rocha-Campos, A. C., 1972, Upper Paleozoic bivalves and gastropods of Brazil and Argentina: A review, *in* 2nd Gondwana Symposium, South Africa, July to August 1970, Proceedings and Papers: Pretoria, Council for Scientific and Industrial Research, p. 605–612.

Rocha-Campos, A. C., 1973, Upper Paleozoic and lower Mesozoic paleogeography, and paleoclimatological and tectonic events in South America, *in* Logan, A., and Hills, L. V., eds., The Permian and Triassic Systems and Their Mutual Boundary: Canadian Society of Petroleum Geologists, Memoir 3, p. 398–424.

Rocha-Campos, A. C., 1979, Late Paleozoic marine fauna of Subandean Bolivia: Affinities, age and paleogeographic implications, *in* Laskar, B., and Raja Rao, C. S., eds., 4th International Gondwana Symposium, Calcutta, 1977: Delhi, Hindustan Publishing Corporation, p. 160–165.

Rocha-Campos, A. C., and Archangelsky, S., 1985 [1987], South America, *in* Martinez-Diaz, C., ed., The Carboniferous of the World, IUGS Publication 20, II—Australia, Indian subcontinent, South Africa, South America, and North Africa: Madrid, Instituto Geológico y Minero de España, p. 175–297.

Romer, A. S., 1973, Permian reptiles, *in* Hallam, A., ed., Atlas of Palaeobiogeography: Amsterdam, Elsevier Publishing Co., p. 159–167.

Romer, A. S., 1975, Intercontinental correlations of Triassic Gondwana vertebrate faunas, *in* Campbell, K. S. W., ed., Gondwana Geology Canberra, National Australian University Press, p. 469–473.

Rong Jia-yu, Boucot, A. J., Su Yang-Zheng, and Strusz, D. L., Biogeographical analysis of Late Silurian brachiopod faunas, chiefly from Asia and Australia: Lethaia, v. 28, p. 39–60.

Rong, J., and Harper, D. A. T., 1988, A global synthesis of the latest Ordovician Hirnantian brachiopod faunas: Transactions of the Royal Society of Edinburgh, Earth Sciences, v. 79, p. 383–402.

Rosen, P. S., 1979, Boulder-barricades in coastal Labrador: Journal of Sedimentary Petrology, v. 49, p. 1113–1124.

Ross, C. A., 1973, Carboniferous Foraminiferida, *in* Hallam, A., ed., Atlas of Palaeobiogeography: Amsterdam, Elsevier Publishing Co., p. 127–132.

Ross, C. A., 1979, Evolution of Fusulinacea (Protozoa) in late Paleozoic space and time, *in* Gray, J., and Boucot, A. J., eds., Historical biogeography, plate tectonics, and the changing environment: Corvallis, Oregon State University Press, p. 215–226.

Ross, J. R. P., 1979, Permian ectoprocts in space and time, *in* Gray, J., and Boucot, A. J., eds., Historical Biogeography, Plate Tectonics, and the Changing Environment: Corvallis, Oregon State University Press, p. 259–276.

Runnegar, B., 1979, Marine fossil invertebrates of Gondwanaland: Palaeogeographic implications, *in* Laskar, B., and Raja Rao, C. S., eds., 4th International Gondwana Symposium, Calcutta, 1977: Delhi, Hindustan Publishing Corporation, v. 1, p. 144–159.

Runnegar, B., and Newell, N. D., 1971, Caspian-like relict molluscan fauna in the South American Permian: American Museum of Natural History Bulletin, v. 146, 66 p.

Rust, I. C., 1973, The evolution of the Paleozoic Cape basin, southern margin of Africa, *in* Nairn, A. E. M., and Stehli, F. G., eds., The Ocean Basins and Margins, v. 1. The South Atlantic: New York, Plenum Publishing Corp., p. 247–276.

Sahni, A., and Khare, S. K., 1973, Additional Eocene mammals from the Subathu Formation of Jammu and Kashmir: Journal of the Palaeontological Society of India, v. 17, p. 31–49.

Sahni, A., and Misra, V. P., 1972, A new species of *Protocetus* (Cetacean) from the middle Eocene of Kutch, western India: Palaeontology, v. 15, p. 490–495.

Sahni, A., Rana, R. S., and Prasad, G. V. R., 1987, New evidence for paleobiogeographic intercontinental Gondwana relationships based on Late Cretaceous–earliest Paleocene coastal faunas from Peninsular India, *in* McKenzie, G. D., ed., Gondwana Six, Stratigraphy, Sedimentology, and Paleontology: American Geophysical Union Geophysical Monograph 41, p. 207–218.

Salamon-Calvi, W., 1933, Epeirophorese, Teil III. Die vordiluvialen Eiszeiten; B, Die Eiszeiten des Karbons und Perms: Akademie der Wissenschaften, Mat.-Phys. Kl., Sitzungbericht, abh. 1, 8 p.

Samylina, V. A., and Yefimova, A. F., 1968, Pervyye nakhodki ranneyurskoy flory v basseyne r. Kolyma: Doklady Akademia Nauk SSSR, v. 179, no. 1, p. 166–168.

Sanchez, T. M., Waisfeld, B., and Benedetto, J. L., 1991, Lithofacies, taphonomy, and brachiopod assemblages in the Silurian of western Argentina: A review of Malvinokaffric Realm communities: Journal of South American Earth Sciences, v. 4, p. 307–329.

Sastri, V. V., and Raiverman, V., 1968, On the basin study programme of Cretaceous-Tertiary sediments of the Cauvery basin, *in* Cretaceous-Tertiary Formations of South India: Geological Society of India Memoir 2, p. 143–152.

Sastri, V. V., Raju, A. T. R., Sinha, R. N., Venkatachala, B. S., and Banerji, R. K., 1977, Biostratigraphy and evolution of the Cauvery basin, India: Geological Society of India Journal, v. 18, p. 355–357.

Saul, J. M., 1967, *Burmeisteria (Digonus) accraensis*, A new homalonotid trilobite from the Devonian of Ghana: Journal of Paleontology, v. 41, p. 1126–1136.

Saul, J. M., Boucot, A. J., and Finks, R. M., 1963, Fauna of the Accraian Series (Devonian of Ghana) including a revision of the gastropod *Plectonotus*: Journal of Paleontology, v. 37, p. 1042–1053.

Saxena, M. N., Gupta, V. J., Meyerhoff, A. A., and Archbold, N. W., 1986, Tectonic and spatial relations between India and Asia since Proterozoic time, *in* Gupta, V. J., ed. Geology of Western Himalayas, Contributions to Himalayan Geology, v. 3: Delhi, Hindustan Publishing Corporation, p. 187–207.

Schaeffer, B., and Patterson, C., 1984, Jurassic fishes from the western United States, with comments on Jurassic fish distribution: American Museum Novitates, no. 2796, p. 1–86.

Scheibnerova, V., 1981, Permian foraminifera of the Sydney basin, *in* Cresswell, M. M., and Vella, P., eds., Gondwana Five: Rotterdam, A. A. Balkema, p. 43–45.

Schlee, J., Behrendt, J. C., and Robb, J. M., 1974, Shallow structure and stratigraphy of Liberian continental margin: American Association of Petroleum Geologists Bulletin, v. 58, p. 708–728.

Schreiber, A., Weippert, D., Wittekindt, H.-P., and Wolfart, R., 1972, Geology and petroleum potentials of central and south Afghanistan: American Association of Petroleum Geologists Bulletin, v. 56, p. 1494–1519.

Schuchert, C., 1935, Correlation of the more important marine Permian sequences: Geological Society of America Bulletin, v. 46, p. 1–46.

Semenoff-Tian-Chansky, P., 1985 [1987], Corals, *in* Martinez-Diaz, C., ed., North Africa, The Carboniferous of the World, IUGS Publication 20, II—Australia, Indian Subcontinent, South Africa, South America, and North Africa: Madrid, Instituto Geológico y Minero de España, p. 374–381.

Sengör, A. M. C., 1984, The Cimmeride Orogenic System and the Tectonics of Eurasia: Geological Society of America Special Paper 195, 82 p.

Seward, A. C., 1897, On the association of *Sigillaria* and *Glossopteris* in South Africa: Quarterly Journal of the Geological Society (London), v. 53, pt. 3, p. 315–340.

Shah, S. C., and Sastry, M. V. A., 1975, Significance of Early Permian marine faunas of Peninsular India, *in* Campbell, K. S. W., ed., Gondwana Geology: Canberra, Australian National University Press, p. 390–395.

Shah, S. C., and Sastry, M. V. A., 1981, Mesozoic history of India and its paleogeographic implications [abs.], *in* Cresswell, M. M., and Vella, P., eds., Gondwana Five: Rotterdam, A. A. Balkema, p. 85.

Shubin, N. H., Crompton, A. W., Sues, H.-D., and Olsen, P. E., 1991, New fossil evidence on the sister-group of mammals and early Mesozoic faunal distributions: Science, v. 251, p. 1063–1065.

Singh, T., 1978, Lower Permian gastropods and bivalves from eastern Himalaya, India: Delhi, Hindustan Publishing Corporation, Recent Researches in Geology 7, p. 276–309.

Singh, T., 1979, Brachiopods from Permian formation of Siang district, Arunachal Pradesh, *in* Gupta, V. J., ed., Upper Palaeozoics of the Himalaya: Delhi, Hindustan Publishing Corporation, Contributions to Himalayan Geology 1, p. 171–188.

Singh, T., 1987, Permian biogeography of the Indian subcontinent with special reference to the marine fauna, *in* McKenzie, G. O., ed., Gondwana Six:

Stratigraphy, Sedimentology, and Paleontology: American Geophysical Union Geophysical Monograph 41, p. 239–249.

Skwarko, W., and Kummel, B., 1974, Marine Triassic molluscs of Australia and Papua New Guinea: Canberra, Bureau of Mineral Resources, Geology and Geophysics, Bulletin 150, p. 111–139.

Smiley, C. J., 1970a, Later Mesozoic flora from Mahan, Pahang, West Malaysia. Pt. 1: Geologic considerations: Geological Society of Malaysia Bulletin, no. 3, p. 77–88.

Smiley, C. J., 1970b, Later Mesozoic flora from Mahan, Pahang, West Malaysia. Pt. 2: Taxonomic considerations: Geological Society of Malaysia Bulletin, no. 3, p. 89–114.

Smiley, C. J., 1974, Analysis of crustal relative stability from some later Paleozoic and Mesozoic floral records, *in* Kahle, C. F., ed., Plate Tectonics — Assessments and reassessments: American Association of Petroleum Geologists Memoir 23, p. 331–360.

Smiley, C. J., 1979, Pre-Tertiary phytogeography and continental drift—Some apparent discrepancies, *in* Gray, J., and Boucot, A. J., eds., Historical biogeography, plate tectonics, and the changing environment: Corvallis, Oregon State University Press, p. 311–319.

Smiley, C. J., 1992, Paleofloras, faunas, and continental drift: some problem areas, *in* Chatterjee, S., and Hotton N., III, eds., New concepts in global tectonics: Lubbock, Texas Tech University Press, p. 241–257.

Smith, A. B., 1988, Late Palaeozoic biogeography of East Asia and palaeontological constraints on plate tectonic reconstructions, *in* Shackleton, R. M., Dewey, J. F., and Windley, B. F., eds., Tectonic evolution of the Himalayas and Tibet, Royal Society of London Philosophical Transactions, A, v. 326, no. 1589, p. 189–227.

Sohn, I. G., and Chatterjee, S., 1979, Freshwater ostracodes from Late Triassic coprolite in central India: Journal of Paleontology, v. 53, p. 578–586.

Spath, L. F., 1928, Revision of the Jurassic cephalopod fauna of Kachh (Cutch): Palaeontologia Indica, n.s., v. IX, Mem. 2, pt. 2, p. 73–133.

Speden, I. G., 1973, Distribution, stratigraphy and stratigraphic relationships of Cretaceous sediments, western Raukumara Peninsula, New Zealand: New Zealand Journal of Geology and Geophysics, v. 16, p. 243–268.

Spence, J., 1957, The geology of part of the Eastern Province of Tanganyika: Geological Survey of Tanganyika Bulletin 28, 62 p.

Spjeldnaes, N., 1982, Palaeoecology of *Ichthyostega* and the origin of the terrestrial vertebrates, *in* Gallitelli, E. M., ed., Paleontologia Come Scienza Geostorica, Modena, Italy, STEM Mucchi Modena Press, p. 323–343.

Spörli, K. B., and Gregory, M. R., 1981, Significance of Tethyan fusulinid limestones of New Zealand, *in* Cresswell, M. M., and Vella, P., eds., Gondwana Five: Rotterdam, A. A. Balkema, p. 223–229.

Spörli, K. B., Aita, Y., and Gibson, G. W., 1989, Juxtaposition of Tethyan and non-Tethyan radiolarian faunas in melanges, Waipapa terrane, North Island, New Zealand: Geology, v. 17, p. 753–756.

Srikantia, S. V., 1981, The lithostratigraphy, sedimentation and structure of Proterozoic-Phanerozoic formations of Spiti basin in the Higher Himalaya of Himachal Pradesh, India, *in* Sinha, A. K., ed., Tectonics–Regional Geology and Biostratigraphy, Contemporary Geoscientific Research in Himalaya, v. I: Dehra Dun, Bishen Singh Mahendra Pal Singh, p. 31–48.

Stankevich, Y. S., 1982, Ostrakody pozdnego mela i osobennosti ikh obitaniya, *in* Martinson, G. G., ed., Mezozoyskiye ozernyye basseyny Mongolii: Leningrad, Izdatesstvo Nauka, p. 145–157.

Stauffer, K. W., 1968a, Silurian-Devonian reef complex near Nowshera, West Pakistan: Geological Society of America Bulletin, v. 79, p. 1331–1350.

Stauffer, K. W., 1968b, Geology of the Khyber Pass, Khyber Agency, West Pakistan: U.S. Geological Survey Project Report No. PK-22, 42 p.

Stauffer, P. H., 1983, Unraveling the mosaic of Paleozoic crustal blocks in Southeast Asia: Geologische Rundschau, bd. 72, p. 1061–1080.

Stauffer, P. H., and Mantajit, N., 1981, Late Palaeaozoic tilloids of Malaya, Thailand and Burma, *in* Hambrey, M. J., and Harland, W. B., eds., Earth's Pre-Pleistocene Glacial Record: Cambridge, Cambridge University Press, p. 331–335.

Stehli, F. G., 1968, Taxonomic diversity gradients in pole location—The recent model, *in* Drake, E. T., ed., Evolution and Environment: New Haven, Connecticut, Yale University Press, p. 163–227.

Stehli, F. G., 1973, Permian brachiopods, *in* Hallam, A., ed., Atlas of Palaeobiogeography: Amsterdam, Elsevier Publishing Co., p. 143–149.

Stevens, G. R., 1973, Jurassic belemnites, *in* Hallam, A., ed., Atlas of Palaeobiogeography: Amsterdam, Elsevier Publishing Co., p. 259–274.

Stevens, G. R., 1980, Southwest Pacific faunal palaeobiogeography in Mesozoic and Cenozoic times: A review: Palaeogeography, Palaeoclimatology, Palaeoecology, v. 31, p. 153–196.

Stöcklin, J., 1984, Orogeny and Tethys evolution in the Middle East. An appraisal of current concepts, *in* Tectonics of Asia, Colloquium 05, 27th International Geological Congress: Moscow, 1984: Izdatestvo Nauka, Reports, v. 5, p. 65–84.

Sues, H.-D., and Olsen, P. E., 1990, Triassic vertebrates of Gondwanan aspect from the Richmond basin of Virginia: Science, v. 249, p. 1020–1023.

Suess, E., 1885, Das Antlitz der Erde, erster Band, zweite Teil: Wien (Vienna), F. Tempsky, p. 311–378.

Suess, E., 1888, Das Antlitz der Erde, zweiter Band: Vienna, F. Tempsky, 703 p.

Suggate, R. P., ed., 1978, The Geology of New Zealand: New Zealand Geological Survey, v. I, 344 p.

Sun Ailin, [Sun, A. L.], 1973a, Permo-Triassic reptiles of Sinkiang: Scientia Sinica, 16, p. 152–156.

Sun Ailin [Sun, A. L.], 1973b, A new species of *Dicynodon* from Sinkiang: Vertebrata Palasiatica, v. 11, p. 52–58.

Suneja, I. J., 1978, Ordovician palaeogeography of northwestern Kashmir (India): Recent Researches in Geology, 4: Delhi, Hindustan Publishing Corporation, p. 323–328.

Talent, J. A., and Mawson, R., 1979, Palaeozoic-Mesozoic biostratigraphy of Pakistan in relation to biogeography and the coalescence of Asia, *in* Farah, A., and DeJong, K. A., eds., Geodynamics of Pakistan: Quetta, Geological Survey of Pakistan, p. 81–102.

Taner, I., and Meyerhoff, A. A., 1990, Petroleum at the roof of the world: The geological evolution of the Tibet (Qinghai-Xizang) Plateau, I: Journal of Petroleum Geology, v. 13, p. 157–178.

Tarlo, L. B. H., 1959, Fossil reptiles from the Panchet beds of India: Nature, v. 183, p. 912–913.

Tarlo, L. B. H., 1967, Triassic reptiles from the shores of Tethys, *in* Adams, C. G., and Ager, D. V., eds., Aspects of Tethyan Biogeography: London Systematics Association Publication 7, p. 103–109.

Taylor, E. L., Taylor, T. N., and Cuneo, N. R., 1992, The present is not the key to the past: A polar forest from the Permian of Antarctica: Science, v. 257, p. 1675–1677.

Teichert, C., 1928, Nachweis palaozoischer Schichten von Sudwest-Neu-Guinea, *in* Nova Guinea, Resultats de l'Expedition Scientifique Neerlandaise a la Novelle-Guinee, v. VI, Geologie, Livr. 3: Leiden, Brill, p. 71–92.

Teichert, C.,1939, The Mesozoic transgressions in Western Australia: Australian Journal of Science, v. 2, no. 3, p. 84–86.

Teichert, C., 1940, Marine Jurassic of East Indian affinities at Broome, northwestern Australia: Royal Society of Western Australia Journal, v. 26 (1939–1940), p. 103–118.

Teichert, C., 1941, Upper Paleozoic of Western Australia: Correlation and Paleogeography: American Association of Petroleum Geologists Bulletin, v. 25, p. 371–415.

Teichert, C., 1942, Marine Upper Jurassic near Derby, north-western Australia: Australian Journal of Science, v. 5, p. 33–34.

Teichert, C., 1943a, The Devonian of Western Australia: A preliminary review, Pt. I: American Journal of Science, v. 241, no. 2, p. 69–94.

Teichert, C., 1943b, The Devonian of Western Australia: A preliminary review, Pt. II: American Journal of Science, v. 241, no. 3, p. 167–184.

Teichert, C., 1949, Permian crinoid *Calceolispongia*: Geological Society of America Memoir 34, 132 p.

Teichert, C., 1950, Discovery of Devonian and Carboniferous rocks in the North-west basin, Western Australia: Australian Journal of Science, v. 12, no. 2, p. 62–65.

Teichert, C., 1951, The marine Permian faunas of Western Australia (an interim review): Palaontologische Zeitschrift, bd. 24, nos. 1/2, p. 76–90.

Teichert, C., 1952, A brief history of the Gondwanaland concept and of the International Gondwana Commission, *in* Teichert C., ed., Symposium sur les séries de Gondwana: 19th International Geological Congress, Alger, 1952, p. 7–12.

Teichert, C., 1958, Australia and Gondwanaland: Geologische Rundschau, bd. 47, p. 562–590.

Teichert, C., 1964, Some biological and paleogeographical factors in the evaluation of ancient climates, *in* Nairn, A. E. M., ed., Problems in Palaeoclimatology: London, Wiley-Interscience, p. 597–600.

Teichert, C., 1966, Stratigraphic nomenclature and correlation of the Permian "Productus limestone," Salt Range, West Pakistan: Geological Survey of Pakistan, Records, 15, pt. 1, p. 1–19.

Teichert, C., 1967, Nature of Permian glacial record, Salt Range and Khisor Range, West Pakistan: Neues Jahrbuch für Geologie und Paläontologie, Abhandlungen, v. 129, p. 167–184.

Teichert, C., 1972, Marine fossil invertebrate faunas of the Gondwana region, *in* 2nd Gondwana Symposium, South Africa, July to August 1970, Proceedings and Papers: Pretoria, Council for Scientific and Industrial Research, p. 125–138.

Teichert, C., 1974, Marine sedimentary environments and their faunas in Gondwana area, *in* Kahle, C. F., ed., Plate Tectonics—Assessments and Reassessments: American Association of Petroleum Geologists Memoir 23, p. 361–394.

Teichert, C., 1981, Permian glaciation in the Salt Range, Pakistan, *in* Hambrey, M. J., and Harland, W. B., eds., Earth's Pre-Pleistocene Glacial Record: Cambridge, Cambridge University Press, p. 278–288.

Teichert, C., 1991, New data on the Permian crinoid family Calceolispongiidae in Western Australia: Royal Society of Western Australia Journal, v. 73, no. 4, p. 113–121.

Teichert, C., and Meyerhoff, A. A., 1972, Continental drift and the marine environment: 24th International Geological Congress: Montreal 1972, Section 7, Paleontology, p. 339–349.

Teichert, C., and Rilett, M., 1974, Revision of Permian Ecca Series cephalopods, Natal, South Africa: University of Kansas Paleontological Contributions Paper 68, p. 1–8.

Teichert, C., and Stauffer, K. W., 1965, Paleozoic reef in Pakistan: Science, v. 150, p. 1287–1288.

Teixeira, C., 1946, Sur la flore fossile de Karroo de Zambesie (Mozambique): Société Géologique de France, Compte Rendu 13, p. 252–254.

Teixeira, C., 1947, Sobre a flora fosil do Karroo da regiao de Tete: Junta de Investigacao Colonial Anais, v. 11, p. 9–28.

Teixeira, C., 1952, La flore fossile du Karroo de la Zambezie et la notion de continent de Gondwana: Congrès pour l'Avancemente d'Études Stratigraphie et Géologie du Carbonifère, 3eme, Heerlen, Compte Rendu, v. 2, p. 627–630.

Termier, G., Termier, H., Lapparent, A. F. de, and Marine, P., 1974, Monographie du Permo-Carbonifère de Wardak (Afghanistan central): Lyons, Faculté des Sciences, Laboratoire de Géologie, Document, Hors Serie 2, 167 p.

Theron, J. N., 1962, The occurrence of fish remains in the Witteberg-Dwyka transition zone: Geological Survey Reports and Annals (Union of South Africa), v. 1, p. 263–267.

Thomson, M. R. A., 1972, Ammonite faunas of south-eastern Alexander Island and their stratigraphical significance, *in* Adie, R. J., ed., Antarctic Geology and Geophysics: International Union of Geological Sciences, ser. B, No. 1: Oslo, Universitetsforlaget, p. 155–160.

Thomson, M. R. A., 1975, New palaeontological and lithological observations on the Lagoupil Formation, north-west Antarctic Peninsula: British Antarctic Survey, Bulletin 41–42, p. 169–185.

Thomson, M. R. A., 1981, Mesozoic ammonite faunas of Antarctica and the break-up of Gondwana, *in* Cresswell, M. M., and Vella, P., eds., Gondwana Five: Rotterdam, A. A. Balkema, p. 269–275.

Thomson, M. R. A., 1983, Late Jurassic ammonites from the Orville Coast, Antarctica, *in* Oliver, R. L., James, P. R., and Jago, J. B., eds., Antarctic Earth Science: Cambridge, Cambridge University Press, p. 315–319.

Thulborn, R. A., 1986, Early Triassic tetrapod faunas of southeastern Gondwana: Alcheringa, v. 10, p. 297–313.

Truswell, E. M., 1980, Permo-Carboniferous palynology of Gondwanaland: progress and problems in the decade to 1980: Bureau of Mineral Resources, Journal of Australian Geology and Geophysics, v. 5, p. 95–111.

Tuan Shuyin, Chen Yey, and Keng Kuochang, 1977, Some Early Cretaceous plants from Lhasa, Tibetan Autonomous Region, China: Acta Botanica Sinica, v. 19, no. 2, p. 114–119.

Turner, S., and Young, G. C., 1992, Thelodont scales from the Middle-Late Devonian Aztec Siltstone, southern Victoria Land, Antarctica: Antarctic Science, v. 4, no. 1, p. 89–105.

Ustritsky, V. I., 1967, O polozhenii severnogo polyusa v pozdnem paleozoye na osnovanii paleontologicheskikh dannyy: Akademia Nauk SSSR, Sibirskoye Otdeleniye, Geologiya i Geofizika, no. 1, p. 25–32.

Ustritskiy, V. I., 1971, Biostratigrafiya verkhnego Paleozoya Arktiki: Leningrad, Izdatelstvo "Nedra," 279 p.

Ustritsky, V. I., 1973, Permian climate, *in* Logan, A., and Hills, L. V., eds., The Permian and Triassic Systems and Their Mutual Boundary: Canadian Society of Petroleum Geologists Memoir 2, p. 733–744.

Ustritskiy, V. I., and Stepanov, D. L., 1976, Paleobiogeografiya i klimat Yevrazii v permi: Paleontologiya, Morskaya geologiya, Moskva, Izdatelstvo "Nauka," p. 103–109.

Ustritskiy, V. I., and Yavshits, G. P., 1971, Middle Carboniferous glaciomarine sediments of the northeastern USSR: Akademia Nauk SSSR, Doklady (English tranls.), American Geological Institute, v. 199, p. 159–161.

Vaslet, D., 1990, Upper Ordovician glacial deposits in Saudi Arabia: Episodes, v. 13, p. 149–161.

Venkatachala, B. S., and Kar, R. K., 1990, Reworked Permian *Dulhuntyspora* in Tertiary sediments of northeastern India: Alcheringa, v. 14, no. 3, p. 177–180.

Verma, K. K., Satsangi, P. P., Srivastava, S., and Mehra, S., 1979, On a plesiosaurian vertebra from the Upper Gondwanas of Kutch, Gujarat, India, *in* Laskar, B., and Raja Rao, C. S., eds., 4th International Gondwana Symposium, Calcutta, 1977: Delhi, Hindustan Publishing Corp., p. 217–220.

Visser, J. N., 1990, The age of the late Palaeozoic glacigene deposits in southern Africa: South African Journal of Geology, v. 93, p. 366–375.

Voskresenskiy, I. A., Kravchenko, K. N., Movshovich, E. B., and Sokolov, B. A., 1971, Ocherk geologii Pakistana: Moscow, Izdatelstvo Nedra, 168 p.

Vozenin-Serra, C., 1984, État de nos connaissances sur les flores de Paléozoïque supérieur et du Mésozoique du sud-est asiatique. Intepretations paléogéographiques: Société Géologique de France Mémoire, n.s., no. 147, p. 169–181.

Wagner, R. H., 1962, On a mixed Cathasia and Gondwana flora from SE Anatolia (Turkey): Congrés pour l'Avancement des Études de Stratigraphie et de Géologie du Carbonifère, 4th, Heerlen 1958: Compte Rendu, t. 3, p. 745–752.

Walcott, C. D., 1889, Stratigraphic position of the *Olenellus* fauna in North America and Europe: American Journal of Science, v. 37, p. 374–392.

Walton, J., 1929, The fossil flora of the Karoo system in the Wankie district, Southern Rhodesia: Geological Survey of Southern Rhodesia Bulletin, v. 15, pt. 2, p. 62–76.

Wang Hongzhen, 1983, On the geotectonic units of Xizang (Tibet) region: Wuhan College of Geology Earth Science Journal, total 19, no. 1, p. 1–8.

Wang Yujing, and Mu Xinan, 1983, Upper Carboniferous and Lower Permian strata in the Gondwana-Tethys province of Xizang (Tibet): Palaeontologia Cathayana, no. 1, p. 411–419.

Warren, A. A., 1982, Australian fossil amphibians, *in* Rich, P. V., and Thompson, E. M., eds., The fossil vertebrate record of Australasia: Clayton, Monash University, p. 145–158.

Warren, A. A., Kool, L., Cleeland, M., Rich, T. H., and Vickers Rich, P., 1991, An Early Cretaceous labyrinthodont: Alcheringa, v. 15, p. 327–332.

Wass, R. E., 1972a, The Permian faunas of eastern and western Australia: A comparison, *in* 2nd Gondwana Symposium, South Africa, July to August 1970, Proceedings and Papers, Pretoria: Council for Scientific and Industrial Research, p. 599–603.

Wass, R. E., 1972b, Permian bryozoa from South Africa: Journal of Paleontology, v. 46, p. 871–873.

Waterhouse, J. B., 1964, Permian stratigraphy and faunas of New Zealand: New Zealand Geological Survey Bulletin, v. 72, 101 p.

Waterhouse, J. B., 1969, A new Permian fauna from New Caledonia, and its relationships to Gondwana and the Tethys, *in* Gondwana stratigraphy, IUGS Symposium [Mar del Plata, Argentina], 1-15 October 1967: Paris UNESCO, Earth Sciences, v. 2, p. 249–272.

Waterhouse, J. B., 1979, Permian rocks of the Kali Gandaki area (Thakkola), north-central Nepal, *in* Gupta, V. J., ed., Upper Palaeozoics of the Himalaya, Contributions to Himalayan Geology, v. 1: Delhi, Hindustan Publishing Corp., p. 194–213.

Waterhouse, J. B., 1982, An Early Permian cool-water fauna from pebbly mudstones in south Thailand: Geological Magazine, v. 119, p. 337–354.

Waterhouse, J. B., 1983, How wide was the Tethys "ocean"?, *in* Gupta, V. J., ed., Stratigraphy and Structure of Kashmir and Ladakh Himalaya, Contributions to Himalayan Geology, v. 2: Delhi, Hindustan Publishing Corp., p. 246–249.

Webers, G. F., 1970, Paleontological investigations in the Ellsworth Mountains, West Antarctica: Antarctic Journal, v. 5, p. 162–163.

Wegener, A., 1929, Die Entstehung der Kontinente und Ozeane (Vierte umgearbeitete Auflage): Friedrich Vieweg & Sohn Aktien-Gesellschaft, Braunschweig, 231 p.

Weippert, D., Wittekindt, H., and Wolfart, R., 1970, Zur geologischen Entwicklung von Zentral-und Sudafghanistan: Beihefte zum Geologischen Jahrbuch, Heft 92, 99 p.

Westermann, G. E. G., 1975, Bajocian ammonoid fauna of Tethyan affinities from the Kambe Limestone Series of Kenya and implication to plate tectonics: Berlin, Newsletter of Stratigraphy, v. 4, p. 23–48.

Whiteman, A. J., 1971, The Geology of the Sudan Republic: Oxford, Clarendon Press, 290 p.

Wilson, E. C., 1990, Permian corals of Bolivia: Journal of Paleontology, v. 64, p. 60–78.

Winkel, R., Ingavat, R., and Helmke, D., 1983, Facies and stratigraphy of the Lower–lower middle Permian strata of the Petchabun fold-belt in central Thailand, *in* Nutalaya, P., ed.-in-chief, Proceedings, Workshop on Stratigraphic Correlation of Thailand and Malaysia: Geological Society of Thailand and Geological Society of Malaysia, p. 293–306.

Wolfart, R., 1970, Fauna, Stratigraphie und Palaeogeographie des Ordoviziums in Afghanistan: Beihefte zum Geologischen Jahrbuch, heft 89, 169 p.

Wolfart, R., and Kursten, M., 1974, Stratigraphie und Palaogeographie des Kambriums im mittleren Sud-Asien (Iran bis Nord-Indien): Hannover, Geologische Jahrbuch, reihe B, heft 8, p. 185–234.

Wongwanich, T., Wyatt, D., Stait, B., and Burrett, C., 1983, The Ordovician system in southern Thailand and northern Malaysia, *in* Nutalaya, P., ed.-in-chief, Proceedings, Workshop on Stratigraphic Correlation of Thailand and Malaysia: Geological Society of Thailand and Geological Society of Malaysia, p. 77–95.

Wood, R. A., Evans, K. R., and Zhuravlev, A. Y., 1992, A new post–Early Cambrian archaeocyath from Antarctica: Geological Magazine, v. 129, pt. 4, p. 491–495.

Wopfner, H., 1991, Permo-Triassic sedimentary basins in Australia and East Africa and their relationship to gondwanic stress pattern, *in* Ulbrich, H., and Rocha Campos, A.C., eds., Gondwana Seven Proceedings: Sao Paulo, Instituto de Geociencias, Universidade do Sao Paulo, p. 133–146.

Wopfner, H., and Kreuser, T., 1986, Evidence for late Palaeozoic glaciation in southern Tanzania: Palaeogeography, Palaeoclimatology, Palaeoecology, v. 56, p. 259–275.

Wright, J. B., Hastings, D. A., Jones, W. B., and Williams, H.R., 1985, Geology and mineral resources of West Africa: London, Allen & Unwin, 187 p.

Xu Ren [Hsu Jen], 1973, Discovery of plant fossils from Mt. Qomolangma region of southern Xizang in China and its significance: Acta Botanica Sinica, v. 15, p. 254–260.

Xu Ren [Hsu Jen], 1976, On the discovery of a *Glossopteris* flora in southern Xizang and its significance in geology and paleogeography: Scientia Geologica Sinica, no. 4, p. 320–335.

Yang Qun and Wang Yu-jing, 1990, A taxonomic study of Upper Jurassic Radiolaria from Rutog County, Xizang (Tibet): Acta Micropalaeontologica Sinica, v. 7, no. 3, p. 195–218.

Yeates, A. N., and 8 others, 1987, The Westralian superbasin: An Australian link with Tethys, *in* McKenzie, K. G., ed., Shallow Tethys 2: Rotterdam, A. A. Balkema, p. 199–213.

Yin Hongfu, 1990, Paleogeographical distribution and stratigraphical range of the Lower Triassic *Claraia*, *Pseudoclaraia* and *Eumorphotis* (Bivalvia): Journal of the China University of Geosciences, v. 1, no. 1, p. 98–110.

Yin Jixiang, Wu Haoruo, Sun Yiyin, Wen Chuanfen, and Liu Tungsheng, 1983, Study on Himalayan stratigraphy in south Xizang (Tibet), *in* Gupta, V. J., ed., Stratigraphy and Structure of Kashmir and Ladakh Himalaya, Contributions to Himalayan Geology, v. 2: Delhi, Hindustan Publishing Corp., p. 59–74.

Young, G. C., 1987, Devonian vertebrates of Gondwana, *in* McKenzie, G. D., ed., Gondwana Six, Stratigraphy, Sedimentology, and Paleontology: American Geophysical Union Geophysical Monograph 41, p. 41–50.

Yuan, P. L., and Young, C. C., 1934, On the discovery of a new *Dicynodon* in Sinkiang: Geological Society of China Bulletin, v. 13, p. 563–573.

Zhang Lujin, 1983a, On the character of Permian microflora in the Junggar basin of Xinjiang: Palaeontologia Cathayana, no. 1, p. 327–365.

Zhang Lujin, 1983b, On the age of the Badaowan Formation in northern Xinjiang: Scientia Sinica, ser. B, v. 26, p. 774–781.

Zhang Lujin, 1990, Permian spores and pollen from Karamay region of Xinjiang, China: Palaeontologia Cathayana, no. 5, p. 181–204.

Zhang Shanzhen, and He Yuanliang, 1985 [1986], Late Palaeozoic palaeophytogeographic provinces in China and their relationships with plate tectonics: Palaeontologia Cathayana, no. 2, p. 77–86.

Zhao Junpu, 1984, Ophiolite and continental suture: Scientia Geologica Sinica, no. 4, p. 359–372.

Zharkov, M. A., 1981, History of Paleozoic Salt Accumulation: New York, Springer-Verlag, 308 p.

Zheng Haixiang, and Zhang Xuanyang, 1990, Is Yarlung Zangbo River belt a suture?, *in* Li Guangcen, Zhou Weiqin, and Nicolas, A., eds., Geology of the Himalayas, v. 2, Papers on Geology: Beijing, Geological Publishing House, p. 209–217.

Zhou Mingzhen, 1981, Vertebrate paleontology in China, 1949-1979, *in* Teichert, C., and Chen Peiji, eds., Paleontology in China, 1979: Geological Society of America Special Paper 187, p. 15–20.

Zimina, V. G., 1967, O *Glossopteris* i *Gangamopteris* iz permskikh otlo zheniy yuzhnogo Primor'ya [*Glossapteris* and *Gangamopteris* from the Permian deposits of the south Maritime Territory]: Paleontologicheski: Zhurnal, no. 2, p. 113–121.

MANUSCRIPT ACCEPTED BY THE SOCIETY JULY 26, 1995

Index

[Italic page numbers indicate major references]

T

U

V

W

X

Y

Z

Typeset in U.S.A. by Johnson Printing, Boulder, Colorado
Printed in U.S.A. by Malloy Lithographing, Inc., Ann Arbor, Michigan